Techniques in Life Science and Biomedicine for the Non-Expert

Series Editor
Alexander E. Kalyuzhny

More information about this series at http://www.springer.com/series/13601

Sylvia Janetzki

Elispot for Rookies (and Experts Too)

 Springer

Sylvia Janetzki
ZellNet Consulting, Inc.
Fort Lee, NJ, USA

ISSN 2367-1114   ISSN 2367-1122   (electronic)
Techniques in Life Science and Biomedicine for the Non-Expert
ISBN 978-3-319-83260-9   ISBN 978-3-319-45295-1   (eBook)
DOI 10.1007/978-3-319-45295-1

Printed on acid-free paper

This Springer imprint is published by Springer Nature
The registered company is Springer International Publishing AG
The registered company address is: Gewerbestrasse 11, 6330 Cham, Switzerland

Preface

When the editor of this book series initially contacted me, my first thought was "No, not another technical publication on Elispot." But in our conversation, I was reminded that this book is not about providing just another technical publication, but an in-depth explanation of the Elispot assay that would allow a newcomer to truly understand the best choices for specific protocol steps, reagents, and materials, and may even give the experienced Elispot user insight into best practices he/she might have not given much thought about (many reasons exist for that, like "too many ready-to-use kits out there," "foolproof" instructions from true or self-acclaimed specialists readily available, and one has "always done it like this"—so why bother). As most of us agree, there is always room for improvement. But one has to find that room based on scientific evidence combined with common sense. Albert Einstein once said: "If you can't explain it simply, you don't understand it well enough," a sensible advice that guided me when studying and later practicing medicine, during my work in translational immunology laboratories and in the world of consulting for the immunomonitoring field. Hence, I would like to focus throughout this book on providing the basic, simple, logical explanations for choices to be made to run the best Elispot possible, supported by over 20 years of work with this assay myself and the shared experience of the many fellow scientists I had and have the pleasure collaborating with. I think about this work as an integration of the collective knowledge of the many excellent immune "monitorers" out there, to whom I dedicate this book.

Fort Lee, NJ   Sylvia Janetzki

Contents

Chapter 1
Overview

1.1 Historic Overview

Elispot is the abbreviation for Enzyme-Linked ImmunoSpot. The name was deduced from the name for another immunological test, ELISA, a technique that is in parts similar to Elispot, and from the fact that spots are the actual readout of the assay. Elispot has been around for many years. It was first described in 1983 for the detection of heat-labile enterotoxin production of bacteria [1], followed by its use for the enumeration of specific antibody-secreting B cells [2]. The assay format was further adapted for the detection of cytokine-secreting cells, and remained generally almost unchanged until today [3]. The assay has gained popularity over the years with a renaissance in the mid-late 1990s, driven by the cancer and HIV research fields [4], which in turn encouraged manufacturers to improve on basic materials and reagents used for the assay to increase its reliability. Their introduction to the field, the availability of newly developed automated Elispot plate readers for the assay analysis [5], and the start of activities focused on assay standardization, harmonization, and validation further boosted Elispot to its current state as one of the most commonly used immunological assays in a wide range of fields and applications, including the research and translational fields of cancer, infectious and autoimmunity-related diseases, transplantation, basic immunology, as well as genetic diseases and AAV-based therapies, and even epidemiology and research on trauma-related injuries of various organs. Over the past decade, an average of 300–400 scientific papers were published each year referring to Elispot. With that, the question arises why Elispot has remained in its popularity and does not seem to fade away into the dungeon of outdated techniques despite the fact that new immunological as well as molecular-biological methods are constantly being introduced that allow scientists to examine the state and/or functions of the immune system and other cells, e.g., polychromatic flow cytometry, HLA multimer-peptide staining, mass cytometry (CyTOF), or T-cell receptor (TCR) sequencing, just to name a few. To answer the question, we need to look at Elispot a little closer.

© Springer International Publishing Switzerland 2016
S. Janetzki, *Elispot for Rookies (and Experts Too)*, Techniques in Life Science and Biomedicine for the Non-Expert, DOI 10.1007/978-3-319-45295-1_1

1.2   Principles of Elispot

The key advantages of Elispot are summarized in Fig. 1.1.

To fully comprehend these advantages, it is essential to understand the principles of the Elispot technology. Elispot allows the enumeration of single cells that secrete a special analyte of interest (analyte = substance that can be measured or "analyzed"). The main prerequisite for Elispot is the availability of an antibody, optimally an antibody pair, that can bind to the analyte. The advantage of using an antibody pair, in which each antibody recognizes a different epitope (the part of an antigen that is actually recognized by antibodies or immune cells) of the analyte, is illustrated in Fig. 1.2.

Generally, if a suspension of single cells and an antibody that recognizes a substance that may be secreted by those cells or a subpopulation thereof exist, then Elispot may potentially be used to enumerate how many cells in that suspension do indeed secrete that substance (e.g., interferon gamma = IFNγ) under a specific condition (e.g., when immune cells are exposed to an antigen). The simplest way to do that is to immobilize the antibody on a suitable solid surface, most commonly on the membrane-based bottom of a 96-well plate (we will examine the membrane choices in detail in Sect. 3.1). The cell suspension is added to the plate wells, and incubated, under various conditions, for a short time (in the range of hours to a few days) to allow cells to release the analyte of interest. Once released, the immobilized antibody on the membrane bottom of those wells captures the analyte via high-affinity non-covalent binding mechanisms. The cell suspension is removed and another antibody that recognizes the analyte is added. The second antibody should preferably recognize a different part (= epitope) of the analyte for optimal binding, as explained above. We now have a sandwich with the membrane on the bottom onto which the first = capture antibody is bound. That antibody binds to one epitope of the analyte, and the second antibody binds to another epitope. The secondary = detection antibody is typically biotinylated (covalently bound to biotin) to allow signal amplification owing to biotin's high affinity to avidin, of which multiple molecules bind to one biotin molecule. Avidin can be easily coupled with an enzyme, which is necessary to convert a soluble chromogenic substrate into a colored precipitate.

Fig. 1.1 Advantages of Elispot.

Elispot

1. Allows to examine cells on the <u>single cell</u> level.

2. Has an outstanding <u>sensitivity</u>.

3. Is a <u>functional</u> assay.

4. Is <u>easy</u> to learn and to adapt.

5. Is a <u>robust</u> assay.

6. Can be adapted for <u>high through-put</u>.

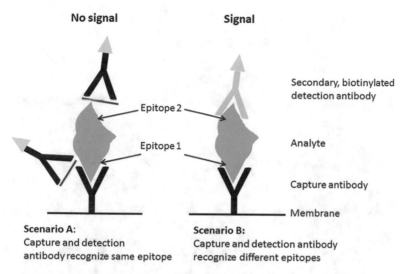

Fig. 1.2 Optimal antibody selection. If the capture and detection antibodies recognize the same epitope of the analyte, detection may be hampered due to occupancy of the epitope by the capture antibody and no or limited binding by the detection antibody (Scenario A). If both antibodies detect different epitopes of the analyte, optimal detection is much more likely (Scenario B). This is a general depiction for optimal antibody choices, which is, conversely, not a black-and-white scenario, and antibodies detecting the same epitope may indeed lead to efficient detection (for example, if the analyte contains repetitive sequences that present the same epitope), and vice versa antibodies that detect different epitopes may not give rise to an optimal signal (for example due to steric hindrance after the capture antibody bound to its epitope). The detailed discussion of under-lying mechanisms is, however, beyond the scope of this book.

Hence, upon stepwise addition of a streptavidin-enzyme complex and suitable sub-strate for the enzyme to act on, the converted substrate precipitates down onto the membrane at the location where the antibody captured the secreted analyte, forming colored spots. The assay principle is also depicted in Fig. 1.3.

The final results of an Elispot experiment are colored spots on a membrane (Fig. 1.4). One spot is the imprint of one cell that secreted the analyte of interest [3]. The morphology of a true spot reflects the secretion kinetics of the analyte. As soon as the analyte is secreted, it is captured by the coating antibody in the immediate sur-rounding of the analyte-secreting cell. The more analyte is secreted (and the faster), the further it can diffuse away from the cell until it is captured, following its concen-tration gradient, until it is entirely bound by the capture antibody choices. We will review resulting implications for the coating procedure in Sect. 3.2. This diffusion and capture process determines the overall appearance of spots, which are defined by a darker center (=high concentration of bound analyte) with fading intensity of color toward the spot periphery (=gradually decreasing concentration of bound analyte). The staining gradient is the most pertinent marker for a true spot caused by active secretion of analyte, and allows its discrimination from other, artificial signals which may be visible on the membrane (discussed in detail in Sect. 6.10).

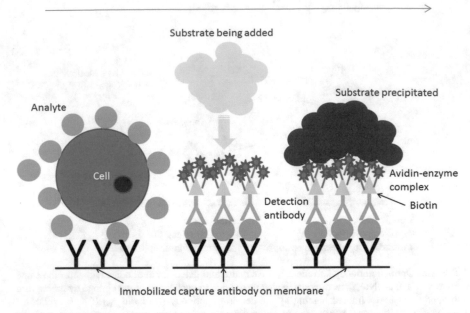

Substrate being added

Substrate precipitated

Analyte

Cell

Avidin-enzyme complex

Detection antibody

Biotin

Immobilized capture antibody on membrane

Fig. 1.3 Principal steps of the Elispot assay.

Fig. 1.4 Eli-"Spots". The image depicts one well with red spots of various sizes. The typical features of true spots can be clearly distinguished: dark centers and fading staining toward the spot periphery. The image was taken with a KS Elispot Reader (Carl Zeiss, Inc., Thornwood, NY).

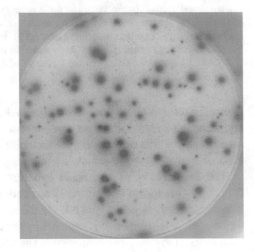

1.3   Modifications

While the overall setup of an Elispot assay, as illustrated in Fig. 1.3, has remained unchanged for the majority of applications, various modifications have been introduced over the past years that led to assay simplification, necessary adaptations for specific situations, and broadening of applications, with the latter one allowing the detection of multiple analytes in one well.

1.3.1 Simplifications

Simplifications of the assay were driven by manufacturers of Elispot kits. They mainly comprise two steps of the assay: (1) the coating step, and (2) the detection step.

1.3.1.1 Coating Step

The coating step takes multiple hours, and is generally performed overnight. The related timeline prohibits the spontaneous setup of Elispot assays, for example when cells become unexpectedly available and are supposed to be freshly tested (without being frozen away), or requires coating to be performed on the weekend to have plates available for a Monday assay. Pre-coated plates have been made available by manufacturers as a convenient alternative. While it appears reasonable to pre-coat plates for future use directly in the lab, it is strongly advised against it. The capture antibody has to be stabilized, a not that trivial procedure. And not all membrane types can be efficiently pre-coated. Hence, self-pre-coated plates which were stored for a prolonged time before use may display inconsistent and suboptimal coating with the capture antibody. Such inconsistency is visible as high variability from well to well, with occasional blank wells, only partially developed wells, and most commonly extremely faint spots which appear as they were washed out. This effect is depicted in Fig. 1.5.

A disadvantage of commercially available pre-coated plates is that they are more costly. Further, the likelihood exists that wells are left unused if an experiment does

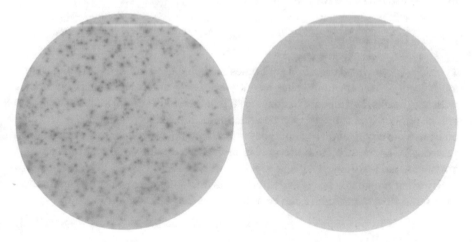

Fig. 1.5 Insufficient coating in self-pre-coated plates. Images from two neighboring wells in a self-pre-coated PVDF plate are shown (no stabilization of capture antibody, stored for more than a week before use), testing the same PBMC for IFNγ secretion upon stimulation with the same antigen. The well on the *right* is strongly impaired for spot development, with spots barely being distinguishable from background, appearing "washed out," due to insufficient amount of antibody bound to the membrane. The well on the *left* shows only minor spot development impairment in the periphery to the *lower right*.

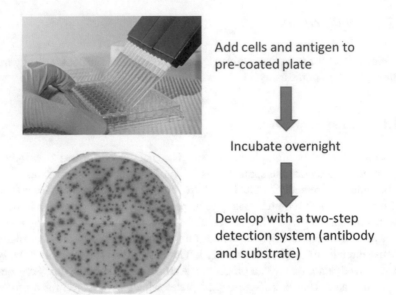

Add cells and antigen to
pre-coated plate

Incubate overnight

Develop with a two-step
detection system (antibody
and substrate)

Fig. 1.6 Simplified Elispot. The main steps for an Elispot experiment using pre-coated plates and an enzyme-conjugated detection antibody are demonstrated.

not require the use of an entire plate, adding to the overall costs. Commercially available pre-coated 8-well strip plates can offer a solution for small-sized experiments. These plates are reviewed in Sect. 3.1.

1.3.1.2   Detection Step

Typically, biotinylated detection antibodies are being used in Elispot to attain signal amplification. Manufacturers have recently been able to label secondary antibodies with a sufficient high amount of enzyme so that the biotin–avidin amplification step can be omitted and substrate can be added after the secondary antibody. The efficient enzyme conjugation of antibodies is challenging, especially using horseradish peroxidase, and hence there are only a few kits currently available offering this simplified methodology.

By using pre-coated plates and directly enzyme-conjugated detection antibodies, the overall Elispot procedure can be as simple as demonstrated in Fig. 1.6.

1.3.2   Adaptations

1.3.2.1   Detection of Specific Antibodies

Adaptations to the outlined general Elispot procedure (Fig. 1.3) are mainly derived from testing B cells for antigen-specific antibody secretion. Historically, antigen-specific immunoglobulins are captured by using the antigen for coating. Secreted antibodies bind directly to the epitope(s) presented by the antigen used for coating, and such are captured on the membrane. The detection follows the traditional way by using biotinylated anti-immunoglobulin. Various challenges are associated with this method, the most prevalent being the effective binding of the antigen to the membrane of choice (nitrocellulose or PVDF, reviewed in Sect. 3.1) and the efficient presentation of the relevant epitope(s) for recognition and binding by the secreted antibodies. This methodology often requires high amounts of antigen for coating in order to obtain well-defined spots (see Sect. 3.2 for more details), a possible restraining factor.

To work around the challenges of efficient antigen binding and epitope presentation as well as high antigen demands, a different detection system for secreted immunoglobulins was developed. Here, all secreted antibodies of one class (e.g., IgG) are captured on the plate by an anti-immunoglobulin (e.g., anti-IgG) antibody. Epitope-specific antibodies are detected by using biotinylated antigen for detection [6]. This method results in higher sensitivity of the assay, with more and better defined spots detected than with the traditional method using antigen for coating. Further, the required amount of antigen per experiment is significantly lower. The drawback is that antigens have to be biotinylated in the lab; however, biotinylation kits are commercially available and the procedure itself is simple and straightforward. Helpful details and advice can be found online at http://www.mabtech.com.

1.3.2.2 Detection of Multiple Analytes in One Well

Dual-Color Elispot

The binding capacity of Elispot membranes is very high, and by far exceeds the amount of antibody used to achieve optimal coating (reviewed in Sects. 3.1 and 6.1). This feature offers the opportunity to coat with two different capture antibodies, specific for different analytes. For the stepwise detection of both analytes, different enzymes and substrates are used that allow the discrimination of spots by color (typically red and blue). If cells secrete both analytes simultaneously, a mixed, purplish color should be detectable. Evidence for successfully establishing a dual-color colorimetric Elispot assay exists [7]. However, one has to overcome multiple challenges, the most prevalent being the setup of conditions that ensures optimal spot color development. Further, strong, large spots of one color may overlay or cover smaller and/or fainter spots of the other color. Depending on the strength of spot development for each analyte, it may be difficult to distinguish dual-stained spots due to only minor color differences to single-colored spots. These issues, which may lead to underestimates of spot numbers for one analyte, have been reviewed and demonstrated in detail elsewhere [8]. An example of a dual-color Elispot is given in Fig. 1.7.

Due to the challenges and limitations of this methodology, it has been mostly abandoned, and replaced with a much more elegant approach that allows for the reliable detection of multiple analytes simultaneously in the same well, the Fluorospot technique.

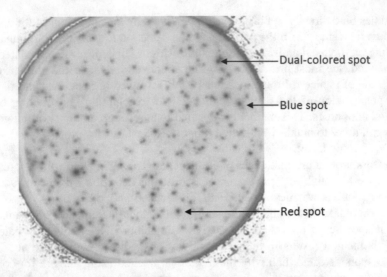

Dual-colored spot

Blue spot

Red spot

Fig. 1.7 Dual-colored colorimetric Elispot. PBMC were tested for IFNγ and IL-2 secretion in the same well upon stimulation with the CEF peptide pool. Detection was performed with different enzymes and substrates, resulting in *blue* and *red spots* for single-secreting cells and purple spots for dual secretors, as indicated by the *arrows*. The image was taken with a Zeiss KS Elispot reader (Thornwood, NY).

Fluorospot

The Fluorospot assay uses a detection system based on fluorophores, while the preceding assay steps are the same as for dual-color enzymatic Elispot. The technique was first introduced in 2003 [9], and later refined [8, 10, 11] including the description of assay validation steps [12]. As of today, two or three capture antibodies with affinity for different analytes are being used for coating, and cells are incubated, and later removed for effective analyte (=spot) detection. Spot detection can be achieved with detection antibodies coupled to a fluorochrome. Signal enhancement steps, e.g., by using a biotinylated secondary antibody, followed by the incubation with streptavidin that is coupled to a fluorochrome, can be useful and is often found in commercially available Fluorospot kits. The most commonly used fluorescent dyes are FITC, Cy3, Cy5, and DAPI (reviewed in Sect. 3.4), in that order. The main requirement for running a Fluorospot assay is the availability of an Elispot reader that is equipped with narrow-band filters for each fluorochrome. Narrow-band filters allow the reliable discrimination of each fluorochrome without encountering bleed-over signals from other fluorochromes used in the assay; and localization algorithms can define the exact position of each spot detected with each filter (=in each channel). The obtained spot coordinates again allow the co-localization of spots in different channels (e.g., in FITC and Cy3) which indicate dual-colored spots (or spots obtained from one cell that secreted both cytokines simultaneously). The advantages of the Fluorospot compared to a single-color Elispot assay are obvious:

1. Less cells are required (e.g., in case of a dual-color Fluorospot half of the amount of cells, and in case of a triple cytokine a third of the amount of cells, is required for testing a sample for the secretion of all cytokines)
2. Less antigen is required (as above, half of the amount for dual-color assays, and a third of the amount for triple-color assays)
3. More information is obtained (the information about cells that secrete more than one cytokine)

In case of a dual-color Fluorospot, three populations are detected: two populations that secrete only one cytokine, and the population that secretes both cytokines simultaneously. In case of a triple-color cytokine Fluorospot, the number of subpopulations detectable increases to seven: three single secretors, three dual secretors, and one triple secretor. An example of the wealth of information assessable is illustrated in Fig. 1.8.

The Fluorospot technique is also applicable for the assessment of B cells [13], allowing the detection of B cells secreting three or even four subclasses of immunoglobulin in one well, correlating with the number of subpopulations detectable, since B cells do not co-secrete different classes of immunoglobulin.

An interesting expansion of the Fluorospot is the introduction of peptide-tagged antigens and fluorophore-coupled anti-tag antibodies that allow, for example, the

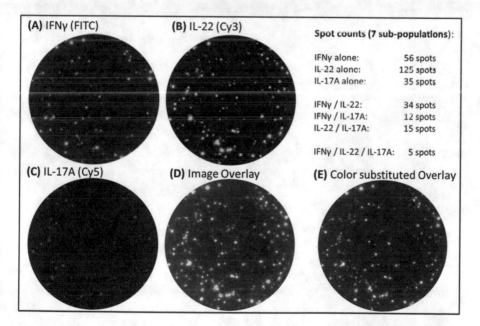

Fig. 1.8 Triple-cytokine Fluorospot assay. PBMC were stimulated with *Candida albicans* extract and tested for IFNγ (FITC), IL-22 (Cy3), and IL-17A (Cy5) secretion. Images for separate fluorophores are depicted in panels (**a**)–(**c**). The image overlay and color-substituted overlay for visualization purposes are shown in panels (**d**) and (**e**), respectively. The overall spot counts for all seven subpopulations are presented. The images were taken with an AID Fluorospot reader system (Strassberg, Germany). The figure was obtained from Janetzki et al. [8].

examination of cross-reactivity levels of antibodies [14]. A different approach to assess cross-reactivity has been introduced in the dengue virus system with fluorophore-coupled serotype-specific detection antibodies [15].

In addition to the assessment of T and B cells with Fluorospot, monocytes have also been successfully evaluated for the release of different cytokines and polyfunctionality [16].

1.4 Summary

The progressing ease of the Elispot assay, combined with its high sensitivity and the ability to assess polyfunctionality on the single-cell level, are important factors that have kept the assay appealing to scientists throughout its existence, despite the increasing number of other assays available to assess the immune system. If the scientific question asked fits in the repertoire of what can be evaluated by Elispot/Fluorospot, then the assay has to be considered or should simply be the go-to assay of choice.

Chapter 2
Important Assay Details

Conducting an Elispot assay is a multistep process. The main steps are:

1. Choosing the correct reagents and materials
2. Preparing the sample
3. Running the assay
4. Acquiring the spot counts
5. Analyzing the data

Before we address the best choices for each step, let's have a look at what can be considered a perfect Elispot assay (Fig. 2.1).

A perfect Elispot has:

- Small, well-defined spots,
- Even spot distribution across the well,
- No artifacts,
- No background staining (elevated staining of the membrane),
- No or very low background reactivity (no or very few spots in negative control wells),
- No false-positive spots,
- Low variability between replicate wells,
- A working positive control,
- A trending control included,

and is

- Repeatable (precise),
- Accurate (measurement is close to the true value).

The next part of this book investigates in detail the necessary steps and choices that enable you to run a "perfect" Elispot.

© Springer International Publishing Switzerland 2016
S. Janetzki, *Elispot for Rookies (and Experts Too)*, Techniques in Life Science
and Biomedicine for the Non-Expert, DOI 10.1007/978-3-319-45295-1_2

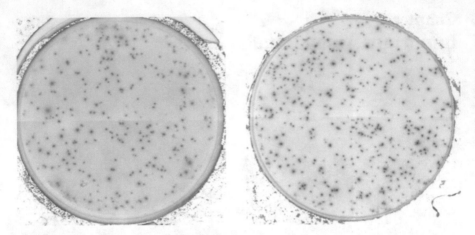

Fig. 2.1 Two wells are depicted from IFNγ Elispot assays testing human peripheral blood mono-nuclear cells (PBMCs) for IFNγ release upon stimulation with the CEF peptide pool. The *blue* and *red* spots were obtained with different enzymatic development systems (reviewed in Sect. 6.9). The membrane bottom of the wells was punched out onto a sealing tape, for optimal visibility of the well periphery. The images were taken with a Zeiss KS Elispot reader system (Thornwood, NY).

Chapter 3
Reagents and Materials

3.1 Elispot Plates

The choice of the Elispot plate can dramatically influence the final results of an Elispot experiment. Membrane-based 96-well Elispot plates are the optimal choice for Elispot. The membrane fibers look like a tightly woven web of polymers that offers a large surface for protein binding (Fig. 3.1).

In general, the protein-binding capacity of membrane-based Elispot plates is more than 100 times higher than that of plastic-bottom plates. The classic Elispot plate choice relies on nitrocellulose ester ("HA" plates). Later, plates based on polyvinylidene fluoride (PVDF)=Immobilon-P ("IP") were introduced. A comprehensive review of the features of both membranes is available elsewhere [17, 18]. Importantly, both membranes differ in their binding mechanism of protein. While the HA membrane is hydrophilic and binds protein via electrostatic mechanisms, the PVDF membrane exposes a high hydrophobicity and binds protein via hydrophobic interactions. However, due to its hydrophobicity, PVDF repels aqueous solutions and hence requires a prewetting step in order to overcome the hydrophobicity and allow optimal protein binding (detailed instructions in Sect. 6.1). Once prewet, PVDF binds protein somewhat more efficiently and, importantly, tighter than its counterpart nitrocellulose ester (e.g., its IgG-binding capacity is about 130 µg per well, compared to 100 µg per well for nitrocellulose). It is hence recommended to work with PVDF plates.

For Fluorospot applications, the PVDF membrane has been further modified to decrease the auto-fluorescence of the regular membrane, which is mainly caused by its porous structure that scatters the light leading to high auto-fluorescence. The much smoother surfaced Immobilon-FL membrane has very low auto-fluorescence levels while retaining the same superior performance characteristics of the regular PVDF membrane. When running Fluorospot, only use plates with the Immobilon-FL membrane.

© Springer International Publishing Switzerland 2016
S. Janetzki, *Elispot for Rookies (and Experts Too)*, Techniques in Life Science
and Biomedicine for the Non-Expert, DOI 10.1007/978-3-319-45295-1_3

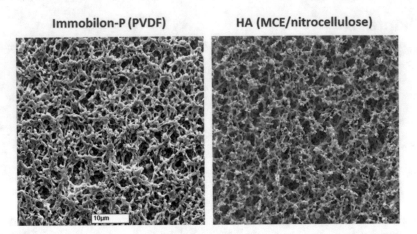

Immobilon-P (PVDF)   **HA (MCE/nitrocellulose)**

Fig. 3.1 Representative cross-sectional scanning electron microscope (SEM) images for the two semi-porous membranes (PVDF and nitrocellulose) most commonly employed in Elispot assays. The high surface area is critical for capture antibody binding and resolution of distinct, sharply defined spots. Images courtesy of MilliporeSigma, Billerica, MA.

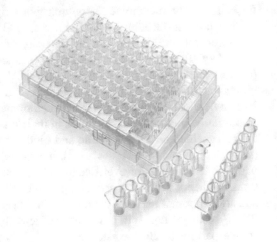

Fig. 3.2 Image of an 8-strip Elispot plate. Constructed in a transparent format, the strips are available in standard PVDF format. The 8-well strips perform equivalently to the standard HTS plates. Image courtesy of MilliporeSigma, Billerica, MA.

One long-standing problem with the 96-well microplates has been the waste of unused wells in small-scale assays such as those occurring in diagnostic analysis of a single patient sample. Recently, an 8-strip version of Elispot plates has been introduced to the market, for experiments that require only a few wells (see Fig. 3.2). The 8-well strips are currently part of Oxford Immunotech's T-SPOT.TB Test, an FDA-approved IFNγ Elispot test kit designed specifically for the diagnosis of tuberculosis infection.

Similarly, 384-well plates are also available, requiring smaller amounts of cells (about three-quarters less compared to 96-well plates) to be plated for optimal setup conditions (reviewed in Sect. 6.6). They perform comparably to the 96-well PVDF plates (Fig. 3.3). Pipetting is possible with regular multichannel pipettors, and their

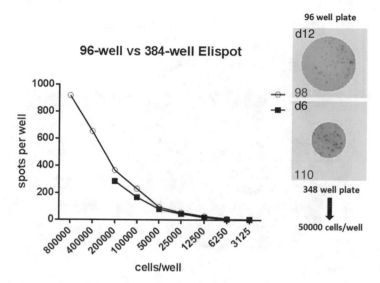

Fig. 3.3 Comparison of IFNγ Elispot results obtained from a 96-well (*open circles*) and a 384-well plate (*black squares*) with the indicated number of MART-1$_{27-35}$-specific CD8+ T cells stimulated with MART-1$_{27-35}$ peptide, demonstrating a tight correlation in spot counts. An example of the well images is given on the *right*. The plates were evaluated with an iSpot Spectrum Reader (AID, Strassberg, Germany). Adding more than 200,000 cells per well to the 384-well plate led to spot confluence and single spots could not be distinguished, hence no results are shown in higher ranges for the 384-well plate. Results, graph, and images courtesy of Drs. Darin Wick, John Webb, and Brad Nelson, Trev and Joyce Deeley Research Centre, British Columbia Cancer Agency, Victoria, BC, Canada.

automated evaluation can be done with most Elispot reader systems currently on the market.

In addition to different types of membranes, plates also have different colored frames, most commonly clear and white or opaque frames. The color does not influence the Elispot results. In general it is easier to track pipetting when using plates with a clear frame. However, an important consideration comes into play when evaluating plates with different colored frames. White framed plates reflect the light back onto the membrane, and an adjustment of camera and exposure settings is required [19]. Because of that reflection issue only clear framed plates are being used for Fluorospot which requires a very strong light source for evaluation.

The original nitrocellulose and PVDF plates were developed for applications other than Elispot. When initially using these plates in Elispot, various challenges became evident which were caused by the early plate design:

- Sporadic lower spot counts in peripheral wells compared to wells located in the center of the plate,
- Spots across the well were not in the same plane (it was not possible to focus spots in the well periphery and well center at the same time) (Fig. 3.4),
- Sporadic leakage (accumulation of liquid in the underdrain below the membranes).

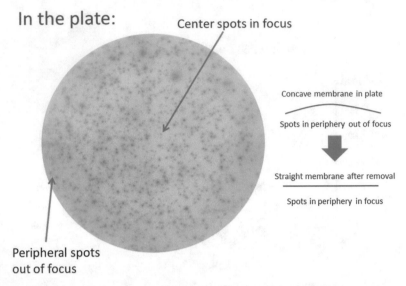

In the plate: Center spots in focus

Concave membrane in plate

Spots in periphery out of focus

Straight membrane after removal

Spots in periphery in focus

Peripheral spots
out of focus

Fig. 3.4 The concave membrane in old format plates ("MA" plates) leads to focus problems during automated evaluation. Membranes had to be punched out and transferred onto a sealing tape for reliable evaluation of the entire well surface. Without membrane removal, the peripheral part of the well was omitted from evaluation. The image was taken with a Zeiss KS Elispot Reader system (Thornwood, NY).

Table 3.1 Plate classification.

	Nitrocellulose membrane	PVDF membrane
Old format plates	MAHA	MAIP
New format plates	MSHA	MSIP

The variability in spot counts was traced back to the open frame construction and resulting temperature gradient in the periphery of plates, while the focus problem was caused by the concave shape of the membrane (the membrane was given room to move and expand as required for other applications). The leakage, an effect of the originally desired filtration function of those plates, was caused by the capillary traction of the filtration spout in the underdrain.

These issues have been addressed by the leading plate manufacturer, MilliporeSigma/Merck. The newly designed plates ("HTS" plates) have a closed-frame design to prevent temperature gradients, a straightened membrane for more even cell/spot distribution, and adjustments of the spout distance between the underdrain and membrane to prevent leakage due to capillary action. The new and old plates can be distinguished by their name. While old plates start with a "MA", new plates start with a "MS" (Table 3.1).

3.2 Antibody Choices

Antibodies for coating and detection belong to the most critical reagent choices for Elispot. It is important to keep in mind that not every antibody (clone) or antibody combination works in Elispot, even though they may work in other techniques. Even the same antibody clone(s), purified and prepared as the final product by different vendors, may perform differently. Hence, it is strongly recommended to work with vendors that select and test their antibodies for performance in Elispot. As outlined in the assay overview earlier on, it is of advantage using two clones (one for capture, the other for detection) that recognize different epitopes of the analyte of interest (see Fig. 1.2). Furthermore, batch-to-batch analysis should be done at the vendor's end and demonstrate low variability, with appropriate documentation available upon request. A few providers for Elispot antibodies and kits (which typically contain also additional reagents for the assay, and possibly plates) have established their position in the field over many years, with Mabtech being a premier provider with strong focus on the Elispot and Fluorospot technologies across their entire product line, followed by Becton Dickinson and R&D Systems, among a few others.

Perhaps the most critical issue related to antibodies is the amount used for coating. Though antibodies carry a hefty price tag, one has to restrain from the temptation to use less antibody for coating than recommended by the manufacturer. A rule of thumb is that about 1 µg of total capture antibody per well (96-well plate) works well for most Elispot assays. The amount correlates with a coating concentration of 10 µg/mL using 100 µL per well. This may vary slightly between different kits and vendors, but rarely strays from the range of 5–15 µg/mL, 100 µl per well. Coating a 384-well plate requires about 0.25 µg total capture antibody per well because the wells are roughly a quarter of the size of a 96-well plate. One exception of the above coating rule is the coating of plates with antigen for B-cell Elispot, as already addressed earlier on (see Sect. 1.3.2.1). Coating concentrations of 10–50 µg/mL and 100 µL per well are often required to obtain acceptable signals = spots.

The logistics behind the coating concentration are demonstrated in Fig. 3.5.

The secondary antibody is typically used in lower concentrations (about 10% of that used for coating), and the influence of its concentration is less dramatic. The lower required working amount typically promotes adherence to the concentrations recommended by the manufacturer.

3.3 Enzymatic Conjugates and Substrates

Almost every commercially available Elispot kit is also equipped with a conjugate of an enzyme required for converting the final substrate into a precipitate, and streptavidin. As a reminder, streptavidin is used to amplify the signal. Detection

The lower the antibody concentration, the larger and fainter the spots

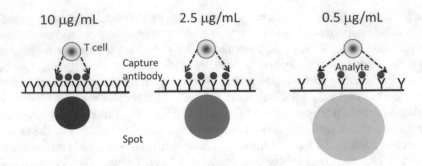

Fig. 3.5 The influence of the concentration of capture antibody on spot appearance. The lower the amount of capture antibody (Y) used for coating, the further apart are the antibodies bound to the membrane (*black line*), the further the analyte (*red dots*) has to diffuse away from the cell secreting it (indicated by *arrows*) in order to be completely captured. This results in larger spots and, due to the fact that less analyte was bound over the same area, also in fainter spots.

Table 3.2 Enzymes, common substrate choices, and resulting spot color.

Alkaline phosphatase (AP)	Horseradish peroxidase (HRP)	Spot color
BCIP/NBT		Purple
	TMB	Blue
	AEC	Red
	DAB	Brown

antibodies are biotinylated for the purpose of amplification, during which multiple streptavidin molecules bind to one biotin molecule.

Two enzyme choices are available: horseradish peroxidase (HRP) and alkaline phosphatase (AP), both acting on different substrates producing different colored spots. There are no clear advantages of one enzyme over the other, and the final choice is typically driven by the preference in substrate use. An overview of enzymes, related substrates, and resulting spot color is given in Table 3.2.

Substrates do differ not only in the color of the spots they produce, but also in their sensitivity. A common substrate for AP is 5-bromo-4-chloro-3-indolyl-phos-phate)/nitro blue tetrazolium (BCIP/NBT) which produces blue–purple spots and exhibits an overall high sensitivity. Different formulations exist (e.g., BCIP alone, BCIP/NBT-Blue, BCIP/NBT Plus), with BCIP/NBT Plus having the highest sensitivity (=precipitating at lowest enzyme concentrations) and hence producing most spots.

There are various substrates available for HRP, with 3,3′,5,5′-Tetramethylbenzidine (TMB) having an exceptional high sensitivity. Two essentially different formula-tions of TMB are available, one with a soluble substrate (as used, for example, in ELISA), and one that allows precipitation (as required for Elispot). Another popular HRP substrate is 3-amino-9-ethylcarbazole (AEC) with a slightly lower sensitivity

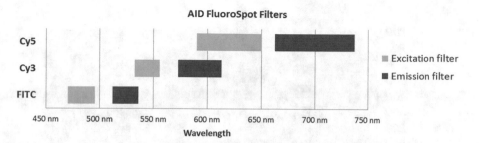

Fig. 3.6 The three commonly used fluorophores and their related narrow-band filters with excitation and emission ranges, as used in the AID Elispot reader (Strassberg, Germany), are depicted. The image was obtained from Janetzki et al. in *Cells* 2015 [8].

than TMB. Lastly, 3,3′-diaminobenzidine (DAB) also works with HRP, forming brown spots. Its sensitivity is lower compared to the other HRP substrates, and it is carcinogenic, and hence should not be used in open applications like Elispot.

Ready-to-use or easy-to-prepare substrate solutions are commercially available from a large variety of vendors, and are also provided by Elispot kit manufacturers.

3.4 Fluorescent Dyes

For the Fluorospot application, no enzymes or substrates are necessary. Instead, the signal is obtained directly from fluorophores coupled to the detection reagents. The choice of commonly used fluorophores is mainly guided by their signal strength and stability. It is further important to prevent bleed-over signals. This can only be achieved by using narrow-band filters integrated in the automated imaging system, which filter the incoming (excitation) and the outgoing (emission) light. Figure 3.6 depicts the three routinely used fluorophores, FITC, Cy3, and Cy5, and their related narrow-band filter ranges.

An excellent tool to investigate the spectra of fluorophores can be found online: http://www.biolegend.com/spectraanalyzer.

Reagents required for the development of a Fluorospot assay are commercially available from the leading antibody vendors, and are included in ready-to-use kits.

3.5 Culture Media

Choosing the right culture medium is a critical step when setting up an Elispot experiment. It has been shown that the choice of medium is the leading cause for variability when different laboratories measure the same sample with Elispot, following their own standard operating procedure (SOP) [20]. As a matter of fact, pretesting the medium/serum is such a critical protocol choice that it is one of the

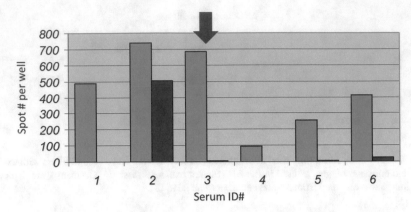

Fig. 3.7 Pretesting results of six different lots of serum. Peripheral blood mononuclear cells (PBMC, 200,000 per well) were tested in Elispot for background reactivity (=cells alone, *purple bars*) and for reactivity against the CEF peptide pool (*blue bars*). The same PBMC were used, but the assay was performed in RPMI supplemented with six different serum lots (ID 1–6). The *red arrow* indicates the serum lot with optimal performance characteristics.

recommendations in the Elispot harmonization guidelines [20]. Most culture media are supplemented with a serum for optimal cell health and function. Various reports exist in published literature about specific sera being superior for immune monitoring in the hands of the reporting laboratory. But the matter of fact is that each lot of serum performs differently, and can potentially suppress immune responses or elicit nonspecific responses (which would result in high background reactivity=many spots in the negative control=the cells-alone condition). Only a few articles elucidate this issue [21, 22]. It is crucial to obtain samples from multiple serum lots for pretesting, in order to choose the one with the best performance characteristics (=low background reactivity, high antigen-specific responses). An example of the outcome of such pretesting is given in Fig. 3.7.

Sources of serum are FCS/FBS (fetal calf or bovine serum), human AB sera, or autologous sera (derived from the donor that is being tested). Human AB serum may give rise to additional concern due to the possible existence of heteroclitic antibodies or other proteins that generate background reactivity or background staining. Because of the challenges in obtaining an optimal serum lot, it was investigated in a multicenter study if serum-free media could be used instead of serum-supplemented media. Such media hold the advantage of having the same defined composition, and hence only need to be pretested once. The study was designed so that the same sets of PBMC were sent to 11 laboratories working with pretested and well-qualified serum-supplemented media. Labs had to test the PBMC side by side with their own culture medium, and four commercially available serum-free media, following their own SOP. It could be convincingly demonstrated that serum-free media performed at least as well as serum-supplemented media, with low background reactivity levels, high spot counts in antigen-stimulated conditions, and high viability and recovery of cells after thawing and even after overnight resting (we will review these steps in Sects. 4.5 and 4.7) [23]. Based on these results it is recommended to consider including serum-free media in pretesting and, based on the results, possibly working with them throughout an Elispot experiment. Some of these media may already be used in the laboratory for other applications, e.g., AIM-V or X-Vivo15. It needs to be clarified that serum-free media should not be supplemented with even low amounts of serum, despite the fact that such practice is described in literature. Even short exposure to low serum amounts can have detrimental effects on cell health and reactivity, as described above, and confounds the entire approach of using serum-free medium in the first place.

3.6 Blocking Buffers

After the Elispot plate has been coated with the capture antibody, a blocking step should be performed to block any potential nonspecific binding to the membrane, which could lead to artificial background issues (also see Sect. 6.2). For this, a buffer containing protein is required. There are two strategies recommended, based on the choice of culture medium used for the assay:

1. If serum-supplemented culture medium is used, then the same medium can be used as blocking buffer. It contains a high concentration of necessary proteins, and prevents the introduction of additional buffers and proteins to the assay.
2. If serum-free culture medium is used, it is recommended to block the plate with phosphate-buffered saline (PBS) without calcium (Ca^{2+}) or magnesium (Mg^{2+}), supplemented with 1% bovine serum albumin (BSA), fraction V

3.7 Washing Buffers

There are different washing buffers needed during the course of an Elispot assay. They are used:

1. For the removal of reagents before cells are being added to the Elispot plate,
2. For the removal of cells from the plate after incubation,
3. For the removal of development reagents,
4. For removal of the substrate to stop spot development.

Because cells need to be incubated under sterile conditions, the buffers used for reagent removal before cells are plated out have to be sterile. All other washing steps can be done under non-sterile conditions. Here is a closer look at buffer choices:

1. Before cells are plated, the prewetting solution (ethanol), unbound capture antibody, and the blocking buffer can be removed by washing the plate with sterile PBS without Ca^{2+} or Mg^{2+}. PBS is preferred to water due to its osmotic and pH features which are supportive in bioassays.

2. The removal of cells after incubation is a crucial washing step. Cells that remain on the membrane lead to impaired spot development (reviewed in detail in Sect. 6.8); also see Figure 6.10. Due to the membrane's porous surface, cells require rigorous washing to be effectively removed. It is recommended to use PBS supplemented with 0.05% polysorbate 20 (e.g., Tween 20). The detergent features of polysorbate 20 help to efficiently remove immunologicals and provide a favorable environment for immune bioassays.
3. The same buffer can also be used for washing steps to remove detection reagents. However, polysorbate should be removed before the final development step with substrate. Therefore, washes with PBS only are recommended before the addition of the substrate.
4. The substrate reaction should be stopped under running tap water. Since the quality of tap water may influence the occurrence of TMB spots (especially ionic compounds), it is recommended to stop the TMB spot development with distilled or deionized water.

3.8 Important Laboratory Supplies

3.8.1 Pipettes

The whole range of serological pipettes is required, from 1 to 20 mL. Further, handheld pipettes are also needed, with sterile and non-sterile pipette tips (allowing the handling of volumes as low as 15 µL/well up to 200 µL/well). Multichannel pipettes are invaluable for addressing the many pipetting steps and wells in an Elispot assay. Select a pipette that allows smooth addition of cells and buffers. Too fast and abrupt addition of cells to the plate can lead to uneven distribution of spots across the membrane, as depicted in Fig. 3.8.

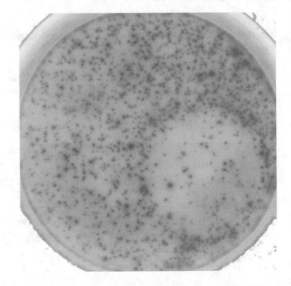

Fig. 3.8 Pipetting effect on spot distribution. The *circle-like area* with limited spots in the lower right part of the well was caused by vigorous addition of cells to the plate, causing cells to be pushed away from their drop point.

Repeat multichannel pipettors can also be used, when the same buffer has to be plated out across the entire plate. These devices aspirate a larger volume of buffer, and then dispense them in defined smaller volume steps, avoiding the repeated aspiration of buffer.

3.8.2   Filters

There are two important filter steps required during an Elispot assay:

1. Filtering the secondary antibody is required to remove aggregates which can cause false-positive spots. Just before adding the diluted antibody to the plate, it is filtered with a low-protein-binding filter, pore size 0.22 μm. Non-sterile syringe filters suffice. Their retention volume is very low, and the aggregate removal is highly efficient.
2. The ready-to-use substrate solution should also be filtered to remove any particle material. Use a non-sterile, low-protein-binding filter, pore size 0.45 μm.

3.8.3   Other Supplies

Common materials used for tissue culture work are required, including 15 and 50 mL tubes, racks, 10 mL syringes, and gloves. Multichannel pipetting reservoirs are recommended for easy pipetting. A centrifuge is required for cell preparation. The main part of the assay has to be done under sterile conditions; hence access to a laminar hood is essential.

3.9   Automated Elispot Readers

Automated imaging systems (=Elispot readers) are available for the enumeration of spots in a plate. These machines take a picture of each well, and based on algorithms and parameters which are user defined within a software program, evaluate these pictures for the occurrence of spots. Spot counts per well are provided in a report format, and information about the spot size and their staining intensity can be obtained as well. Although this process sounds easy and straightforward (counting spots on a membrane), there exists a multitude of factors that can influence the final spot counts. As a matter of fact, spot counting has been shown to be one of the main sources for variability in Elispot [19, 20]. It is highly subjective, and requires a deep understanding of how to sort out faint and strong, "good" and "bad" signals visible on the membrane [19, 24]. (We need to remind ourselves here that Elispot is run in an open environment, and is subject to many disturbances from "internal" components related to the sample plated out into a well and external interferences related to the plate handling.) We will review the analysis of Elispot plates in detail in Sect. 6.10. The main Elispot reader systems are available from A.EL.VIS (Hannover, Germany), AID (Strassberg, Germany), Biosys (Miami, FL), CTL (Cleveland, OH), and Zeiss (Thornwood, NY). The latter has recently stopped its reader production. All manufacturers provide a variety of machines that differ in hardware and software features from basic machines that allow the evaluation of enzymatic Elispot plates in 96-well plates to highly complex machines also allowing the analysis of Fluorospot assays and other format plates (e.g., 384-well plates) and assays. Readers with robotic features are also available for high-throughput plate analysis.

Chapter 4
Sample Preparation

4.1 Overview

Elispot is a functional assay. Hence it is important to meet sample preparation requirements that support proper function of cells. It is therefore necessary to isolate mononuclear cells from whole blood. Further isolation of subpopulations of cells can be performed. In general, Elispot can be performed on

- Peripheral blood mononuclear cell (PBMC)
- Tumor-infiltrating lymphocytes (TILs)
- Cells from lymph nodes (LN)
- Spleen cells (splc)
- Single-cell suspensions obtained from specific tissue
- Cells from pulmonary lavage
- Cultured cell lines (non-adherent; in suspension) and clones
- Other single cells in suspension

    The key factors related to the sample influencing the outcome of an Elispot assay are:

1. Viability of cells
2. Degree of apoptosis in the cell population
3. Degree of contamination with "unwanted" cells
4. Functional state of cells
5. Available co-stimulation
6. Cell density
7. Effectiveness of antigen presentation
8. Contamination with cell and tissue debris

    Depending on the population of cells assessed, some factors are more prominent in their influence than others. In the following section we will investigate important protocol steps and related parameters that are critical for obtaining an optimal sample.

© Springer International Publishing Switzerland 2016
S. Janetzki, *Elispot for Rookies (and Experts Too)*, Techniques in Life Science and Biomedicine for the Non-Expert, DOI 10.1007/978-3-319-45295-1_4

4.2   Whole Blood and PBMC

An effective Elispot assay to evaluate T- and/or B-cell responses cannot be performed on whole blood [25]. PBMC have to be isolated before testing. There are excellent recommendations available for purifying and working with PBMC [26, 27]. What follows is a summary of the most important aspects of obtaining suitable PBMC preparations for functional evaluation:

4.2.1   Blood Draw and Anticoagulants

Blood is obtained via venipuncture [28], and should always be regarded as potentially hazardous, and hence all proper precautions have to be taken. To obtain a suspension of single mononuclear cells, whole blood needs to be anticoagulated (to prevent clotting). While blood anticoagulated with any of the three commonly used anticoagulants, ethylenediamine tetraacetic acid (EDTA), heparin, and citrate, has been shown to be suitable for preparations of PBMC and can lead to similar performance in Elispot [29], evidence exists that blood stored in EDTA is less stable and may deteriorate faster [30], while blood collected in heparin, specifically in lithium–heparin, may have some advantage in functional studies [25]. Based on the many reports available, it is recommended to work with heparinized blood.

4.2.2   PBMC Isolation

The cell population of interest for the majority of functional immune studies is the PBMC population. It contains the lymphocytes (T, B, and NK cells) and monocytes. These cells are characterized by a round nucleus. Cells without a nucleus (erythrocytes, platelets) and cells with multi-lobular nuclei (granulocytes) need to be removed for effectively studying the above immune cells. The most common procedure is the Ficoll density gradient centrifugation [31]. Depending on the cells' density, they migrate through the Ficoll medium during centrifugation and form layers of cells. Red blood cells aggregate when in contact with the Ficoll and hence sediment fast to the bottom of the centrifuge tube. Enhanced by contact with Ficoll, granulocytes, which have a relatively high density, penetrate through the Ficoll and accumulate below it, right above the red blood cells. On the other hand, the density of mononuclear cells is not high enough to allow them do so as well. They accumulate right between the interface of Ficoll and the original blood sample and form a clearly defined, cloudy-appearing band – the PBMC ring. Above that ring the plasma phase with the low-density platelets can be found. The graph in Fig. 4.1 illustrates the distribution of cells and Ficoll before and after centrifugation.

The success of the PBMC isolation is dependent on various factors. Importantly, Ficoll and blood have to be added to the tube without the two of them mixing.

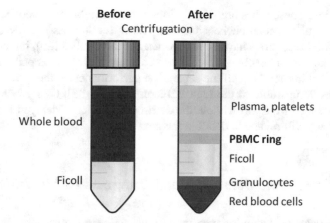

Fig. 4.1 Distribution of cells and Ficoll before and after centrifugation.

This can be achieved by underlaying the blood with Ficoll, or overlaying the Ficoll with blood during a very slow and careful pipetting approach. This rather hideous protocol step can be avoided nowadays by using tubes with a porous firm disk (Leucosep or Accuspin tubes) that can be prefilled with Ficoll (below the disk). Once blood is available, it can be fast and easily added to the tube (above disk) without the risk of mixing it with Ficoll. The time saved using this method compared with the layering method is significant, while the yield of PBMC is comparable to the original method [32].

Another factor influencing the success of PBMC isolation is the temperature of the Ficoll, which changes its density accordingly. It has to be room temperature (optimally 18–20°C), since red blood cells efficiently aggregate at contact with the Ficoll at ambient temperature, but not at lower temperatures. Lower temperatures prevent the red blood cells and granulocytes to permeate the higher density Ficoll, hence leading to their contamination of PBMC. On the other hand, Ficoll at higher temperatures and therefore lower density will allow mononuclear cells to enter, leading to low recovery rates. In line with these facts, the centrifugation should also be done at room temperature.

It is recommended to dilute blood 1:1 with PBS (at room temperature), to avoid trapping of mononuclear cells at high cell densities by red blood cell aggregates. However, when using Leucosep or Accuspin tubes, a large enough sediment of red blood cells is required to push the PBMC ring above the porous disk. It hence seems appropriate to avoid blood dilution for tubes with a separation barrier as found in these tubes or CPT tubes (reviewed below) [26].

Lastly, the recommended centrifugation speed has to be followed closely to allow cells to effectively separate, and brakes need to be turned off when centrifugation stops, to leave cell layers undisturbed.

Tubes also exist that combine blood draw and PBMC isolation. They are called CPT tubes, and contain an anticoagulant (heparin or citrate) as well as a density gradient liquid and a gel that allows movement and penetration of cells upon centrifugation. These tubes are slightly longer than regular blood collection tubes, and hence require special buckets and centrifuges for successful PBMC isolation. Similar PBMC recoveries have been reported for these tubes [29], but also slightly higher contamination levels with red blood cells and platelets (as per the manufacturer's product insert). Overall the separation performance of these tubes has been described as comparable to the regular Ficoll method [33]. CPT tubes do offer a solution for blood transport without the detrimental effect of delayed PBMC isolation, which we will review in detail in the next chapter.

4.2.3   Time Frame Between Blood Draw and PBMC Isolation

Very few aspects of the entire experiment including sample preparation have such profound effect on the final Elispot results as the time lapse between blood draw and PBMC isolation. Two groups simultaneously provided evidence that PBMC isolation needs to be done within 8 hours of blood draw in order to obtain optimal Elispot responses [29, 34]. But reports already exist since the early 1980s about the negative effects of delayed blood processing [35].

The eminent question is this: What actually causes the observed decline in functionality? It has been found that the main culprits are granulocytes [32], which get activated after blood draw. Here we look at the related mechanisms that influence the Elispot results:

1. Once granulocytes get activated they change their buoyancy profile and with that cannot be effectively isolated from the PBMC during the Ficoll gradient centrifugation [36, 37]. Instead, they are isolated, frozen, thawed, processed, and counted as if they were PBMC. This overestimation leads to a relative dilution of PBMC and decrease in spot numbers correlating with the degree of granulocyte contamination (e.g., 20% granulocyte contamination=20% less PBMC added=20% less spots). Many granulocytes will die during the freezing and thawing process, and add to the clumping effect often observed during thawing (reviewed in Sect. 4.5). But the effect of granulocytes does not stop here.

2. Once granulocytes are plated out into the Elispot plate, they may bind to the Fc portion of the capture antibody via their Fc receptor [38]. With that, they get further activated (the activation effect will be looked at under point 3). But since they are now bound to the antibody, they stick to the membrane and lead to a physical disruption of spots, as demonstrated in Fig. 4.2.

3. As already mentioned in points 1 and 2, granulocytes get activated upon blood draw. This leads to the release of hydrogen peroxide [36], and the activation of arginase, which in turn leads to a downregulation of the CD3 zeta chain expression and with that to an interference of the T cell receptor (TCR) signal transduction [36, 37, 39, 40]. These mechanisms are ultimately responsible for the suppression of functionality as observed in functional T cell assays like Elispot after delayed blood processing.

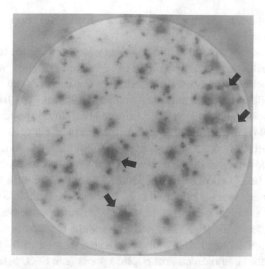

Fig. 4.2 Well image with *blue large spots* disintegrated by *white* "spots" indicative of granulocyte contamination. Granulocytes bound to the Fc portion of the capture antibody and led to *"white"* ghost spots due to their occupancy of the membrane during the initial development steps. The effect is only visible on membranes with spots due to the inhibition of regular spot formation. *Red arrows* indicate some prominent areas of spot disintegration due to granulocyte contamination.

Clearly, the question evolves whether anything can be done to either avoid granulocyte contamination or counteract the effects of their activation. Obviously, a fast isolation of PBMC puts this discussion to an end. Being able to isolate PBMC within a few hours after blood draw keeps the contamination with granulocytes and the effects of their activation in check. McKenna and colleagues had a closer look at the degree of granulocyte contamination after various scenarios following blood draw [37]. They obtained multiple tubes of blood from a donor. First, they measured the degree of granulocyte contamination when they isolated PBMC right after blood was drawn. The next tube was evaluated for granulocyte contamination of PBMC after 8-h storage at ambient (room) temperature, and was found to have about double as many granulocytes. When blood was stored for 24 h at ambient temperature, the contamination degree increased about tenfold. If blood was stored at 4 °C for 24 h, the contamination level increased close to a 100-fold. But interestingly, if blood was diluted 1:1 with either plain PBS or RPMI after blood draw, and then stored for 24 h at ambient temperature, the degree of granulocyte contamination in the following PBMC preparation was low and comparable to that of PBMC preparations obtained after 8-h blood storage at room temperature. The dilution of blood clearly counteracted the granulocyte activation, as confirmed also by others [32]. Hence, if blood cannot be processed within 8 h, it would be an option to train the personnel at the blood-drawing site to dilute the blood under sterile conditions. Blood can be, for example, diluted into 50 mL tubes prefilled with PBS, the cap properly closed to prevent leakage, and transported/shipped to the laboratory at ambient temperature, where PBMC isolation will occur at a delayed point of time.

The approach has been recommended by the T-Cell Workshop Committee of the Immunology of Diabetes Society [26].

A logical alternative method would be to remove granulocytes from whole blood. While a kit (T-Cell Xtend) was developed to cross-link granulocytes with red blood cells for effective removal from PBMC during the Ficoll isolation step, proof of effectiveness has been sparse, and only shown for PBMC not frozen and thawed before use in Elispot [41, 42]. Two large, unpublished studies performed at Viracore-IBT Laboratories (Lee's Summit, MO) addressed the effect of this kit on the functionality of T cells obtained after delayed isolation, followed by freezing and thawing. The following details were provided for this book as courtesy by Viracore-IBT. Blood from healthy donors was collected in lithium–heparin tubes, and PBMC were isolated either immediately or after 24 h of storage (mimicking "ambient temperature" transport conditions). Isolation of PBMC after 24 h was performed with or without the use of T-cell Xtend, and cells were then frozen. PBMC samples (freshly isolated and isolated after 24 h) were tested simultaneously by IFNγ Elispot for responses to two common control peptide pools (see Sect. 5.1.2), the CEF peptide pool [43] eliciting MHC class I-restricted responses and the CMV pp65 overlapping peptide pool [44], which can elicit MHC class I- and class II-restricted responses. In both independent studies, there was a distinct decrease of responsiveness observed to both peptide pools after 24-h storage of whole blood before processing in the majority of conditions (donors, peptide pool, and operator) without the addition of T-cell Xtend, as measured by IFNγ Elispot, compared to immediately isolated PBMC. Both studies failed to demonstrate a positive effect of T-Cell Xtend on the functionality of PBMC (increase or recovery of peptide pool-specific responses to levels of freshly isolated PBMC) after delayed PBMC isolation.

This comes as no surprise, because granulocytes already underwent activation before Ficolling, and suppressed T cell functionality before PBMC isolation, as described in point 3 above. The only way to circumvent that is the early granulocyte removal right after blood draw. Ferromagnetic beads are available that can be used in whole blood. They only require a short incubation, followed by the placement of the blood tube into a magnet, and transfer of the granulocyte-free blood into a new tube. Sterile working conditions are required, along with training of personnel. Unfortunately, no published reports are available at the time this book was written about their successful use including the examination of any potential nonspecific stimulation effects of such beads.

A possible work-around can be achieved by using CPT tubes, which provide a closed system for blood draw and PBMC isolation, as described above. Once blood was drawn, tubes need to be spun down within 2 h following exactly the manufacturer's instructions. After centrifugation, PBMC, platelets, and plasma are contained above the polyester gel barrier. Granulocytes are located below that barrier, successfully separated and blocked from asserting their immune-suppressive effects on the lymphocytes. The tube(s) should be inverted a few times to mix PBMC and plasma. They can then be transported/shipped to the laboratory where further processing takes place. The success of this method is dependent on the centrifugation being done shortly after blood draw and the ambient temperature of tubes (to avoid red blood cell aggregates and trapping of mononuclear cells), the use of appropriate centrifuges and buckets (horizontal rotor with swing-out heads, correct buckets for the longer than usual CPT tubes), and correct centrifugation speed.

Lastly, if whole blood has to be shipped overnight, any temperature drop below ambient temperature has to be avoided (see McKenna study for effects of cold temperatures on granulocyte contamination as described earlier; [37]). By monitoring the temperature in shipment containers, Olson et al. have shown that packages that were shipped as ambient temperature shipment are often exposed to cold outside temperatures which ultimately affect the temperature within the shipping box [45]. Based on this study, the group is recommending the use of reusable gel packs within the shipment container, pre-warmed to 37 °C in winter months, and at room temperature in summer months.

4.3 Immune Cells Obtained from Tissue

Immune cells can be obtained from a variety of tissues. Most commonly, these comprise tumor samples for the preparation of tumor infiltration lymphocytes (TILs), lymph nodes, and spleens (from mice) for an according sample preparation. Independent of tissue, the sample preparation requires the dissociation of the tissue by mechanical means, which can be combined with enzymatic degradation of the extracellular matrix. Whatever method used, the goal is to obtain a single-cell suspension free of remaining tissue. Protocols including video demonstrations for the isolation of immune cells from tumor samples have been published [46]. Miltenyi Biotec offers an entire line of products for a standardized cell isolation approach from tissue including tissue storage solution, dissociation tubes and kits, a dissociator, and cell strainers and filters.

When isolating spleen cells, it is recommended to remove red blood cells by hypotonic lysis before using the obtained single-cell suspension (for a detailed protocol, see [21]).

4.4 Freezing of Cells

Real-time testing of freshly isolated cells, while desirable, is only feasible in small research experiments and maybe in early, well in advanced planned and controlled small-scale clinical studies. With the inherent variability in single-cell functional immune assays as Elispot looming (we will review the variability issue in Chap. 8), it is strongly recommended to batch test multiple samples, e.g., samples from the same donor or patient obtained at different time points along a certain disease pathway, treatment, or vaccination schedule. While freezing obviously exerts stress on cells and hence may impact their viability and functionality, optimized and standardized freezing and thawing techniques can ascertain similar functional activity of frozen cells compared to fresh cells [47–49].

Cells are typically frozen in the presence of dimethyl sulfoxide (DMSO) which serves as a cryoprotectant by preventing the formation of ice crystals in cells that would lead to their lysis and death. The optimal cooling rate is 1 °C/min, which can be achieved with three methods:

1. Automated controlled-rate freezers;
2. Mr. Frosty™, a plastic container that relies on a similar cooling rate of isopropanol at −80 °C;
3. CoolCell® containers, which provide a controlled 1 °C/min freezing rate in −80 °C freezers due to the employment of a thermo-conductive alloy core and special outer insulation.

Automated freezers clearly provide the easiest and straightforward freezing method, but may be cost prohibitive for some laboratories. A much cheaper and currently most commonly used alternative is the method utilizing Mr. Frosty. It depends on the use of isopropanol. It is cheap and effective. CoolCell has entered the market more recently. It relies, as Mr. Frosty, on a reusable container. Various advantages of CoolCell make this method an attractive go-to solution for freezing, like the standardized controlled freezing rate, no need of isopropanol or any other reagents, and the touchability and reusability of the freezing container right after removing it from −80 °C.

Freezing protocol details are available in published literature [27, 50].

In addition to the optimal cooling rate and the presence of DMSO, another factor is crucial for the optimal functionality of cells after freezing. This factor pertains to the freezing medium used. Historically, freezing media consist of a mixture of high-percentage serum and DMSO. As already described in Sect. 3.5, even short exposure to serum with suppressive or nonspecific stimulatory features can have detrimental effects on the assay outcome (e.g., missed response detection or high background reactivity). It has been shown that the same samples frozen in media supplemented with different lots of sera can exhibit different reactivity levels in Elispot within a laboratory and across laboratories [27]. Similar to the study designed and executed to investigate the performance of serum-supplemented and serum-free media in Elispot [23], a large study was performed to investigate the applicability of serum-free freezing media [51]. PBMC were frozen in a freezing medium supplemented with a highly qualified serum, as well as in a commercially available serum-free freezing medium and a self-made serum-free freezing medium based on human serum albumin (HSA). Those cells were sent to 31 laboratories in 13 countries for Elispot testing for reactivity against 3 peptide antigens (which were also provided) with the labs' own SOP. The study revealed that the average cell recovery and cell viability after thawing were at least as good or higher when using serum-free freezing media. Similarly, the cell loss after overnight resting was in average better (=lower) when cells were frozen in freezing-free media. Most importantly, the overall background reactivity and overall response detection were better in samples frozen in serum-free media. Because of the excellent study design and its size it can be concluded with high confidence that "serum-free freezing media support high cell quality and excellent Elispot assay performance across a wide variety of different assay protocols" (see paper title). Other reports underline the suitability of HSA-based serum-free freezing media for PBMC [52].

It is important to stress that the above studies refer to the assessment of PBMC for IFNγ secretion. However, evidence exists that certain cell subsets may exhibit decreased recovery after freezing, or impacted functionality. A summary of these observations is provided elsewhere [26], and the interested reader is advised to visit the article, since a detailed review of the topic of susceptibility of specific subsets of lymphocytes to freezing is beyond the scope of this book.

Serum-free freezing media are commercially available from various sources. They can also be self-made using HSA (typically 10–12.5% HSA, 10% DMSO).

While cells are frozen at temperatures between −70 and −80 °C, they should be transferred into liquid nitrogen shortly thereafter for long-term storage to prevent loss of functionality [48, 53]. The loss may be attributed to the fact that cells are not yet entirely frozen at −80 °C; hence some metabolic activity still exists. Liquid nitrogen storage offers two options: the total immersion of cells in liquid nitrogen, and the storage of cells in the vapor phase of liquid nitrogen. The difference between these two options is their temperature, which is −196 °C in liquid nitrogen, and −150 °C in its vapor phase. What is the difference between these three temperature points for freezing cells? The difference is related to the glass transition temperature of water, which is at −137 °C. Below that temperature molecular movement ceases and biological activity is completely suspended. Hence both liquid nitrogen storage options are sufficient for PBMC storage.

It has been shown that CD4+ T cells may exhibit an increased apoptosis rate after long-term (>6 months) storage even in liquid nitrogen [54] and decreased functionality, while CD8+ T cells seemed not to be affected. Apoptosis of important cells may be inhibited by incorporating apoptosis inhibitors (e.g., zVAD-fmk) to the cryopreservation and culture media used after thawing. This is until today, however, not a common procedure for functional T cell assays; hence the evidence is limited about the effectiveness of their use. A review about targeting cryopreservation-induced cell death is available [55].

4.5 Thawing of Cells

While freezing is typically performed slowly, thawing of cells is done fast at 37 °C. It is absolutely crucial to use pre-warmed thawing medium, and to add it slowly to the cells while there is still a small residue of ice left in the vial [49, 50, 52]. These recommendations are based on the assumption that the opposite would induce an osmotic shock to the cells. The most commonly used method is the thawing of vials in a 37 °C water bath. It takes about 150–160 s to melt one 1 mL vial of frozen PBMC. The ice completely melts when it reaches 10 °C. Vials should not be shaken during the thawing process to prevent microcrystals from poking holes into the cell membrane. A standardized thawing process can be achieved with the use of the ThawSTAR® system, which has a similar thawing profile compared to the water bath, while controlling the thawing temperature and alarming the user when it is time to remove the vials and add thawing medium.

A common observation after thawing is clumping of cells. Once these clumps have formed, they cannot be resuspended. These clumps are caused by DNA released from dying cells, which acts as a super-sticky matrix. To prevent clumping, it is recommended to add a DNase to the thawing medium. For PBMC, the use of Benzonase® has been shown to be most beneficial [48, 49]. Its sole effect is the increase in the recovery of cells due to the removal of DNA and hence prevention of clumping. It has no influence on the final Elispot results. Benzonase has to be removed by washing the cells with Benzonase-free culture medium before they can be utilized further.

Freezing and thawing of murine splenocytes are possible and follow similar logistics as described for PBMC [56, 57].

4.6   Cell Counting and Viability

A common approach to determine the cell concentration and viability is the trypan blue exclusion method using a hemacytometer, a standard manual cell counting method in many laboratories. Trypan blue is a dye that is excluded from living cells with an intact cell membrane. However, it enters dead cells which have an impaired cell membrane that allows the dye to pass into the cell and to stain the cytosol blue. Cells are visually examined under a microscope, where cells stained blue are counted as dead cells [58]. The hemacytometer is a cell glass counting chamber with a specifically designed grid that allows the exact calculation of the cell concentration. Although simple and straightforward, the method is prone to variability in final cell counts, depending on the competency of the operator. The operator has to be trained in the use of the hemacytometer, preparation of cells, the setup and loading of the counting chamber, and the inclusion of the correct cells in the final counts as per their location on the grid. Further, the operator has to be able to distinguish lymphocytes and monocytes from platelets, red blood cells, and debris. Lastly, the correct blue staining intensity for dead cells has to be recognized. Appropriately trained scientists can obtain accurate cell and viability counts (live/dead) with this procedure. Over the past decade automated cell counters, based on the trypan blue exclusion method, have been introduced to the market [59], contributing to efficiency and reproducibility in cell counting.

One shortcoming of trypan blue is that it does not stain apoptotic cells, cells that have entered the nonreversible path of programmed cell death, and hence may lead to an overestimation of living cells in a sample [60, 61]. Apoptosis can be induced by various sorts of stress exerted on cells during isolation, freezing, and thawing. While apoptotic cells are counted as living cells with trypan blue, they actually only dilute the living cell fraction with nonresponsive cells that cannot produce spots [62]. Furthermore, it has been shown that they may impact processing of complex antigens (e.g., lysate) that require processing and presentation by professional antigen-

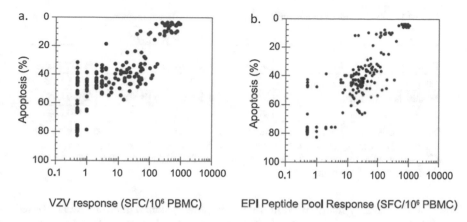

Fig. 4.3 Relationship between Elispot responses obtained from PBMC of the same origin, stored under optimal and suboptimal conditions, and degree of apoptosis. Apoptosis was assessed with the Guava Nexin assay. Healthy cells are found on the *top* of the graph (close to 0% apoptosis as indicated on the y-axis). Spot counts (expressed as spots per 1 million PBMC) can be found on the x-axis on a logarithmic scale (!), thus cells exhibiting the strongest responses against either CD4-restricted epitopes (**a**) or CD8-restricted epitopes (**b**) are located on the right side within each graph. The figure was adapted from the original figure in Smith et al., Clin Vaccine Immunol 14(5): 527-537 (2007) [63].

presenting cells before they can be recognized by T cells, thus further weakening the detectable Elispot response [62]. In a study conducted by Smith et al., that aimed at establishing quality acceptance criteria for previously frozen PBMC for further use in Elispot, samples were stored at optimal and suboptimal conditions and their degree of induced apoptosis determined, followed by testing in an IFNγ Elispot assay for CD4- and CD8-restricted responses (Fig. 4.3) [63]. With increasing degree of apoptotic cells in the sample, the detectable Elispot response decreased significantly. Overall, the study determined that the degree of apoptosis was the most reliable acceptance criterion for cell health and functionality. In the authors' hands, an apoptosis degree of <18% post-thaw was necessary to obtain reliable spot counts.

The obvious question is this: How can the degree of apoptosis be determined if trypan blue does not allow the discrimination between living and apoptotic cells? In a large Elispot harmonization study, in which participants obtained the same cells and antigens, which they had to test in an IFNγ Elispot assay with their own SOP, evidence crystallized that laboratories that counted the apoptotic cells reported overall lower viability of the shipped PBMC samples after thawing [20]. Most importantly, this group had an overall better Elispot performance, likely due to the fact that the PBMC plated in the Elispot assay were adjusted to a more accurate amount of living cells [64]. The assessment of apoptotic cells belongs now to one of the harmonization guidelines for Elispot [20].

One option for assessing apoptosis is the use of flow cytometric assays. However, a standard flow cytometry approach may be time, labor, and cost prohibitive. Over the past decade, systems equipped with a micro-capillary design (avoiding

Table 4.1 Comparison of precision in cell counting for different methods.

	Cell concentration data		Viability	
Platform	Average % CV	% CV range	Average % CV	% CV range
Image-based automated counter	9.2%	1.2–23.3%	3.7%	0.8–12.1%
Manual hemacytometer	6.3%	0.5–15.3%	4.5%	0.5–9.2%
Muse® cell analyzer	4.0%	0.3–8.8%	2.2%	0.4–5.6%

Data are based on triplicate measurements of 30 cellular samples from suspension and adherent cell lines at multiple concentrations and viabilities. The coefficient of variation (CV) is defined as the ratio of the standard deviation (SD) to the mean. The table was provided as courtesy from MilliporeSigma, Billerica, MA.

sheath fluid that is typically used in traditional flow systems) and fluorescent-based detection have been developed, which together with mix-and-read optimized kits offer fast and precise options for cell and viability counting (as well as the assessment of other parameters). The first system that was also used by the subgroup of laboratories in the Elispot harmonization studies mentioned above was the Guava® PCA system. Smaller, benchtop models have been introduced since then, for example the Muse® system, allowing simple, effortless operation and analysis even for a novice to flow cytometry. Their advantage in comparison to manual and other automated counting methods is highlighted in Table 4.1.

The imminent significance of the apoptotic cell fraction is demonstrated in Fig. 4.4. While healthy PBMC preparations exhibit a low degree of apoptotic cells post-thaw (typically between 4 and 6%, Sample 1), impaired cell preparations can contain a large number of cells that have entered programmed cell death but which are not dead yet (Sample 2), and hence cannot be detected by trypan blue exclusion.

4.7   Overnight Resting

It was suggested early on that resting of cells after thawing allows apoptotic cells to die. With cell debris washed out afterwards the rested cell population contains overall more living for counting and Elispot testing [21]. Resting was also shown to improve the reactivity of effector cells without increasing background reactivity [22]. Further evidence for a favorable effect of overnight resting was obtained in the named harmonization studies by the Cancer Immunotherapy Consortium (CIC), in which laboratories that rested cells before Elispot testing had an overall better Elispot performance compared to laboratories that did not [20]. The same observations were made in similar European efforts [65]. Overnight resting of cells has since been included in the Elispot harmonization guidelines.

Much evidence has evolved recently that there is more to resting than removal of apoptotic cells, an effect that was also confirmed via flow cytometric analysis [66].

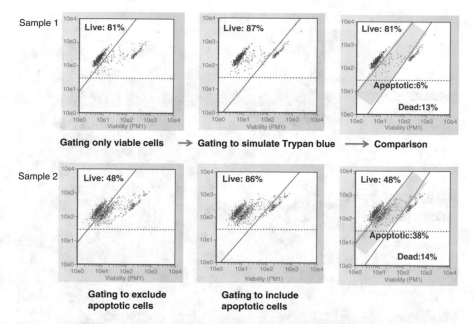

Fig. 4.4 Two PBMC samples (Sample 1 and Sample 2) were analyzed with MilliporeSigma's Count and Viability assay using the Guava® easyCyte flow cytometer. The live cell population is located on the *left* of each graph, while the dead population is located on the *right*. Signals between both populations represent apoptotic cells. Gating on the *left* demonstrates the correct gating approach excluding apoptotic cells. The *center* graphs simulate trypan blue counting, with apoptotic cells included in the live cell population. The *right* graphs depict the size of the apoptotic cell fraction (highlighted in *green*). The graph was provided as a courtesy of MilliporeSigma, Billerica, MA

The same group also demonstrated changes in the quantity as well as the quality of T cells responding to viral antigens. The monofunctional T cell fraction decreased upon overnight resting, while the fraction of multifunctional T cells as well as their sensitivity to antigen increased. This increase was not caused by a proliferation of cells during resting, as tetramer staining revealed the same number of TCR-specific T cells before and after resting.

The work by Römer and colleagues provided profound insights into the mechanisms responsible for the improved functionality [67]. To cut a long and exciting story short, Römer (for CD4 cells) and later Wegner (for CD8 cells) [68] demonstrate how T cells lose their sensitivity to antigen and hence their responsiveness when entering circulation, and how high-density pre-culture can reset them to their much more responsive (they call it "tissue-like") state they were in before entering circulation. The fact that T cells lose their primed stage when entering the circulation had already been revealed earlier in the murine model [69]. Responsible for that loss is the lack of weak TCR signals from HLA scanning provided by tissue-dependent interactions [70]. The simple and easy-to-follow RESTORE protocol

| | No resting | 16 hours resting | 22 hours resting |

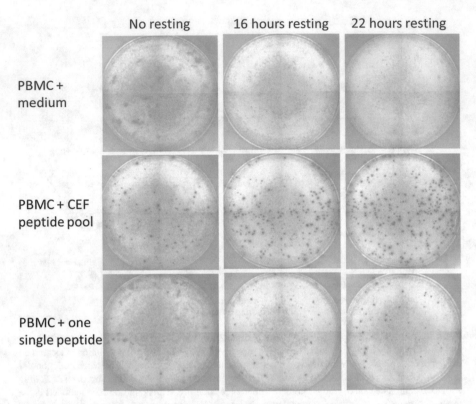

PBMC + medium

PBMC + CEF peptide pool

PBMC + one single peptide

Fig. 4.5 Frozen PBMC were thawed and tested in an IFNγ Elispot assay for background reactivity (PBMC+ medium, *upper panel*), and for reactivity against the CEF peptide pool (containing 32 peptides of 8–11 amino acid length from the cytomegalovirus (CMV), Epstein–Barr virus (EBV), and influenza virus [43], *middle panel*), and a single influenza peptide (*lower panel*). Cells were either tested right after thawing (*left column*) or after resting for 16 h (*middle column*) and 22 h (*right column*). Images were taken with a KS Elispot reader system (Zeiss, Thornwood, USA). Images were provided as courtesy by Drs. Clemencia Pinilla, Valeria Judkowski, Alcinette Buying, and Nazila Sabri from Torrey Pines Institute for Molecular Studies (San Diego, CA).

provided by Wegner provides conditions to the cells mimicking tissue conditions, thus increasing the T cell functionality [68]. The interested reader is strongly advised to study both of these excellent publications (by Römer and Wegner).

Recently, the improvement of IFNγ Elispot performance following overnight resting of frozen PBMC samples was confirmed in a study supported by rigorous statistical analysis [71]. The effect is also clearly demonstrated in Fig. 4.5. The decrease of small, speckle-like artifacts, which are a hallmark of dying cells during the Elispot assay, is dramatic the longer cells are rested (correlating with the fact that less apoptotic cells are added to the Elispot assay, the longer the resting period lasts). The increase in spot numbers after resting is self-evident, and does not require any statistical testing.

Overnight resting has also been demonstrated to have a profound positive impact on antibody-dependent cell-mediated cytotoxicity (ADCC) and natural killer (NK) cell activity in ^{51}Cr-release and CD107a assays [72]. The authors show that frozen and overnight-rested PBMC had higher ADCC and NK activity in both assays when compared to fresh PBMC, while frozen, but none rested PBMC exhibited significantly lower ADCC and NK activities than fresh PBMC. These facts are relevant here since Elispot can also be used to assess NK cell activity.

The importance of using the correct overnight resting protocol needs to be stressed. It is essential to provide tissue-like conditions = high density for sufficient cell-to-cell contact. Sterile 50 mL conical tubes are well suited for resting. The cap needs to be loosened for gas exchange when placing them in the incubator. Alternatively, 50 mL conical tubes with vent caps (a membrane that allows aseptic free gas exchange) may be used. Avoid using tissue culture flasks or tissue-culture-coated plates. They do not provide the needed high-density culture conditions, and lead to the loss of monocytes owing to adherence. Detailed resting protocols can be obtained in published literature [68, 71].

Overnight resting has been widely investigated and used in cells that were previously frozen. With the recently gained insight into the mechanisms of restoring PBMC functionality by resetting them to a tissue-like state it has to be hypothesized that the effect of resting on cells freshly isolated from the circulation should be similar to that of frozen and thawed cells. No study exists that systematically investigates this hypothesis. However, the Wegner publication gives evidence that this is indeed the case [68]. They describe the reactivity of fresh and frozen cells without resting as comparable. Cells undergoing the RESTORE procedure clearly surpass the functional capacity of non-rested cells, freshly isolated or frozen.

4.8   Isolation of Subpopulations of Cells

Isolating subpopulations of lymphocytes may be of interest to either define the responding cell populations (e.g., CD4+ or CD8+ T cells) or enrich cells to obtain sufficiently strong signals in Elispot. For example, the average percentage of CD8+ T cells in PBMC preparations is about 20%. When plating 200,000 PBMC into a well, only 40,000 cells are CD8 positive. If only 1/5,000 CD8+ T cells secrete cytokine, the spot count would be 8 spots per well (which is typically close to the limit of detection which we will review in Chap. 7). However, after isolating CD8+ T cells, potentially 200,000 purified cells can be plated (5x more than originally), and with that the detectable spot number per well may increase fivefold (in our case up to 40 spots). But with the removal of all other PBMC components, antigen-presenting cells and cells providing essential co-stimulation are also missing. These missing factors have to be addressed with the addition of the antigenic stimulus in order to obtain sufficient CD8+ T cell responses. Similar logistics apply for work with splenocytes or cells obtained from other tissues. Important guidelines for plating cells (cell number, stimulation) will be reviewed in detail in Sect. 6.6.

Cell subsets can be isolated with magnetic beads which are typically coupled with an antibody against a cell surface marker (e.g., against CD4 or CD8). After exposure of cells to such beads, the cell suspension is placed into a magnet and cells not bound to the beads are washed off. Some slightly different methodologies exist, with the MACS, Dynal, and EasySep methods currently being used most widely. Some methods require the detachment of cells from the magnetic beads after cell isolation to prevent hindering cell-to-cell contact in downstream testing, or nonspecific stimulation. Positive selection refers to the purification of cells of interest with magnetic beads directed against their surface markers. Negative selection refers to the removal of unwanted cells with magnetic beads directed against their surface markers (here wanted cells are washed off from cells that are held back in the magnet due to their binding to the magnetic beads). Detailed protocols are provided by the manufacturers.

4.9 Expansion of Cells

One of the outstanding strengths of Elispot is its ability to identify cells in very low frequencies, substantially surpassing the capability of flow cytometry-based methods [73]. However, frequencies of antigen-specific T cells may be below the detection limit of Elispot [74, 75]. A solution to that dilemma may be provided by the in vitro expansion of cells during which cells are stimulated in the presence of antigen(s) and cytokines for multiple days [75]. These conditions can lead to a proliferation of antigen-specific cells and hence to an increase of their frequency, and a detectable response in following Elispot testing. There are, however, multiple challenges related to the in vitro stimulation (IVS) of cells:

1. The expansion of antigen-specific cells depends on their proliferative potential (which may partially be influenced by exerted stress on cells during cell processing, granulocyte contamination, etc. as reviewed above), and hence cells may or may not expand or expand only at an unknown rate.
2. Elispot results obtained from cells undergoing IVS do not allow an estimation of the precursor frequency in the original sample. These first two facts have important implications when comparing responses from two time points. If both time points reveal an existing response, it cannot be reliably determined if the response indeed increased or decreased in strength between both time points.
3. The culture conditions may lead to nonspecific stimulation of other cells, potentially elevating background reactivity levels to a degree at which existing antigen-specific responses cannot be distinguished, a common problem in IVS [76].
4. A sufficiently controlled IVS/Elispot experiment requires multiple controls (e.g., for specificity a control should be included in which PBMC are only stimulated with cytokine, but not antigen). Further, including an external negative control (stimulating cells which are not expected to have a response against the antigens of interest) adds another layer of confidence to obtained Elispot testing results.

5. IVS stimulation requires a high number of cells, owing to the controls and general setup conditions required and the fact that over the course of the stimulation non-proliferating cells may die.
6. Because of the many factors that can influence the final results obtained by Elispot, it is imperative for obtaining any confidence in the data to repeat each experiment involving an IVS.

The challenges of IVS were recently addressed in a multicenter study organized by the European Association of Cancer Immunotherapy, CIMT. The same samples were sent out to various laboratories, which expanded them in vitro following their own SOP, followed by testing with Elispot and multimer staining [76]. Overall results were non-reproducible, especially for samples with low or non-detectable responses before IVS. IVS protocols applied differed widely by lengths of stimulation, type and amount of cytokines added, and time point of cytokine addition, among other parameters. Based on the initial study outcome the IVS protocol was harmonized, reducing the variability between laboratories and leading to better response detection among participating centers. The interested reader is advised to consider following the published harmonization protocol, if an expansion of antigen-specific cells is truly demanded [76].

It was recently discussed that an Elispot with and without IVS may quantify different cell populations, namely the effector memory cell population via Elispot without IVS, and the central memory cell population via Elispot with IVS, providing overall different information on T-cell-mediated immunity [77].

Keeping the high sensitivity of Elispot and the challenges associated with IVS in mind, it is recommended to run the Elispot assay without a preceding IVS, except if central memory T cells are investigated.

Chapter 5
Planning the Experiment

5.1 Controls

A good Elispot experiment starts with a well-thought-through plan of execution. The quality of an experiment and its results are determined by the controls it includes. Here is a summary of important controls:

5.1.1 Negative Controls

Cells alone + medium. This is the most important control of a typical Elispot experiment. It controls the background reactivity level of the tested sample. Spot counts in antigen-stimulated wells are compared to spot counts in the cells-alone control to determine the response status (see Chap. 7).

An exception is the B-cell Elispot. Since the antigen is already included in the coating or detection procedure, no such negative control as for cytokine Elispot exists. This seems less of a problem considering that cytokine secretion by a T-cell can be caused by a variety of factors, including nonspecific stimulation or the transition to an active effector cell state in vivo by different antigens the donor was exposed to (e.g., concurrent infection). In contrast, a given B-cell only produces antibodies in response to one specific epitope, and hence requires appropriate activation. A B-cell Elispot detects only those B-cells that secrete the epitope-specific antibody (if the antigen is included in coating or detection). A negative control could include antigens which should not be recognized by the donor, and/or a donor that should not recognize the test antigen.

Cells + negative control antigen. This is a control of less importance compared to the cells-alone control. The negative control antigen is typically an antigen that the cohorts of donors or animals in the study are not expected to have a response against. In the past, various studies included HIV-related peptides or peptide pools for

© Springer International Publishing Switzerland 2016
S. Janetzki, *Elispot for Rookies (and Experts Too)*, Techniques in Life Science and Biomedicine for the Non-Expert, DOI 10.1007/978-3-319-45295-1_5

Fig. 5.1 False-positive spots in an IFNγ Elispot assay. The well was incubated with medium only. Some of the spots are large and have a comet tail, which is a distinct, but not always present feature. The image was taken with a Zeiss KS Elispot reader system (Thornwood, NY).

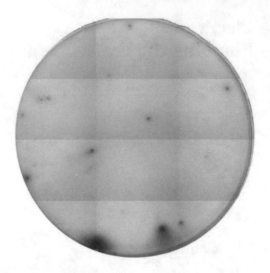

donors assumed to be HIV negative. It is, however, advised not to include such control since it has to be perceived as HIV testing of the donor, likely without consent. Negative control peptide pools are available, e.g., obtained from actin or HLA. As explained above, using a negative antigen is a feasible and good control for B-cell Elispot.

Cells + "empty" antigen presenters. As reviewed in Sects. 6.4.3 and 6.6, an Elispot assay may be set up with separate antigen-presenting cells (APC) which are loaded with the antigen(s) of interest. In such case it is imperative to test the reactivity of cells against the APC per se, not loaded with antigen.

Medium only. Wells are coated and developed like any other well, but do not receive cells or stimulants, only medium. This controls the occurrence rate of false-positive spots, which are caused by aggregates of biotinylated development reagents. False-positive spots cannot per se be distinguished from real spots. They may appear larger than real spots, and they may show a "comet tail" (similar to a shooting star), as depicted in Fig. 5.1. See Sect. 6.9 for further details on how to avoid the occurrence of false-positive spots.

5.1.2 Positive Controls

Cells + nonspecific stimulation: Cells are incubated with a nonspecific stimulator to check if the experiment worked in general and if cells were overall responsive. Various stimulators can be used dependent on the cell population of interest:

(a) Phytohemagglutinin-L (PHA-L), a lectin derived from plants that cross-links lymphocytes and polyclonally activates T-cells: It works well for human PBMC and for a wide range of cytokines, including IFNγ, IL-2, IL-5, granzyme B, and many more. Response to PHA has also been recommended as a quality acceptance criterion for PBMC [63]. A commonly used final concentration in Elispot is 10 µg/mL.

(b) Concanavalin A (ConA), also a plant-derived lectin that polyclonally activates T-cells: It is commonly used for mouse experiments testing splenocytes with a final concentration of 2–4 µg/mL. ConA also works well for nonhuman primate PBMC at slightly higher concentrations of 6–10 µg/mL.

(c) Phorbol 12-myristate 13-acetate and ionomycin (PMA/Iono) are both used in combination to stimulate the intracellular production of various cytokines, e.g., IFNγ, IL-2, IL-4, and perforin. While PMA activates the signal transduction enzyme protein kinase C, ionomycin acts as an ionophore that increases the intracellular levels of Ca^{2+}. Commonly used final concentrations in Elispot are 50 ng/mL for PMA and 1 µg/mL for ionomycin.

(d) Staphylococcal enterotoxin B (SEB), a superantigen that activates a significant fraction of T-cells by cross-linking MHC class II molecules with T-cell receptors, causing the release of cytokines: It is typically used at a final concentration of 10 ng/mL. SEB belongs to biological toxins on the Federal Select Agent Program list; hence its use, storage, disposal, inventory, and reporting must meet state and federal requirements.

(e) Anti-CD3 (monoclonal antibody directed against the ε chain of CD3), a polyclonal activator of T-cells that is commonly used for human PBMC at a final concentration of 100 ng/mL.

(f) R848 + IL-2, a polyclonal activator of B-cells for IgM, IgG, and IgA secretion: A recent publication revealed the superior effectiveness of R848 in combination with IL-2 to turn memory B-cells into an activated state [6]. R848 is a Toll-like receptor (TLR) 7/8 agonist. Pre-stimulation of B-cells is carried out for 24–72 h with 1 µg/mL R848 and 10 ng/mL IL-2.

(g) Anti-CD40 (monoclonal antibody against CD40) + IL-4, a polyclonal activator of B-cells for IgE secretion: Pre-stimulation of B-cells is carried out for 5 days with 1 µg/mL anti-CD40 and 30 ng/mL IL-4.

(h) Lipopolysaccharide (LPS), a TLR4 agonist that stimulates monocytes and dendritic cells (DCs) to release a variety of cytokines including GM-CSF and TNF: It is typically used at a final concentration of 100 ng/mL.

Cells + antigen-specific stimulation. Such a control serves as an internal control for samples obtained from a donor. It typically comprises a collection of antigens (peptides) against which a donor is likely to exhibit a reactivity. The most commonly used control is the CEF peptide pool [43], a collection of MHC class I-restricted peptides obtained from CMV, EBV, and Influenza virus. The peptides are well characterized, and bind to common HLA class I alleles. A large percentage of the Caucasian population has been shown to exhibit a response against that peptide pool, which is stable over time. The original pool of 23 peptides was extended to 32 peptides to increase the HLA allele coverage. MHC class II-restricted control peptide pools (derived from tetanus) are also available, alone or in combination with class I-restricted peptides. Another commonly used peptide control is the CMV pp65 peptide pool that contains 15 mers overlapping by 11 amino acids, which span the entire length of the protein [44]. Such internal control fulfills the purpose to check the responsiveness of a given human PBMC sample at any time point tested. Large discrepancies in reactivity levels of samples from the same donor obtained from different time points should trigger further investigational steps. Also, when a sample is tested repeatedly, the internal control provides information about the precision of the assay (also see Chap. 8).

5.1.3 External Controls

An external sample control serves as a trending control for Elispot performance [4]. It typically comprises 1–2 reference samples (a "positive" = with reactivity against the standard antigen, and a "negative" sample without reactivity against the standard antigen) that are tested against medium only, against a standard antigen (e.g., the CEF peptide pool) and a mitogen each time an experiment is run. Reference samples are samples obtained from the same donor and time point, which are frozen away in multiple aliquots. Such reference samples can be obtained commercially, but may be cost prohibitive for many laboratories. Laboratories may be able to produce their own reference samples with an IRB-approved protocol in place. An attractive option is the use of TCR-engineered reference samples [78]. A-do-it-yourself kit allows the transduction of PBMC with selected T-cell receptor (TCR) RNA. Positively transduced lymphocytes are spiked back into original PBMC samples which can be frozen and used upon demand. Such reference samples provide control over the assay performance over time (trending). They can also control the inter-laboratory reproducibility.

It is advised to also consider establishing a reference sample for mouse experiments.

5.2 Plate Layout

An Elispot experiment requires testing of multiple conditions, often for multiple samples. A logical and easy-to-follow plate layout is therefore important to avoid pipetting mistakes (e.g., cross-contamination of wells with wrong cells or stimulators) and to ensure an easy workflow. Grouping of samples and antigens is recommended, as demonstrated in Fig. 5.2. Each condition should be tested in at least triplicates, if the available cell number allows. For statistical reasons, it should be attempted to plate the negative control (typically cells + medium) in six replicates (reviewed in detail in Chap. 7).

If more than one plate is used in an experiment, each plate needs to be clearly labeled on the plate cover and the plate frame. New-format MS plates (see Sect. 3.1) can also be barcoded. For multi-plate experiments, the medium-only control and reference sample(s) only need to be included on one plate.

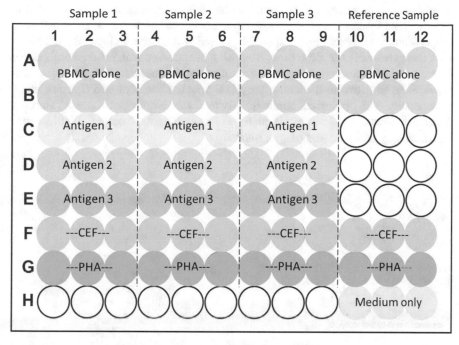

Fig. 5.2 A recommended plate layout for testing of three PBMC samples against three experimental antigens. The template also includes the negative control (cells + medium) in six replicates, an internal control (CEF), a positive mitogen control (PHA), an external trending control (reference sample), and a control for the occurrence of false-positive spots (medium only).

5.3 Timing

An Elispot experiment comprises three main steps, the preparation of cells, the Elispot assay, and the data acquisition and analysis.

A conventional IFNγ Elispot assay takes 3 days if plates are self-coated. On day 1, the plate is coated and incubated overnight, and on day 2 the plate membrane is blocked and cells and stimulants are plated out, followed by 16–24-h incubation. On day 3 cells and stimulants are removed and spots are developed. The assay only takes 2 days if pre-coated plates are utilized.

Some other cytokines may require a longer incubation, e.g., 40–48 h for IL-4 and 48–72 h for IL-17. The length of the assay increases accordingly.

If cells undergo an in vitro expansion, the start of the IVS has to be carefully planned depending on its length, to allow the assay to be conducted during the work week.

If cells are rested before plating, they need to be thawed (if frozen) and prepared on day 1, when the plate is coated. Cells without resting (e.g., freshly prepared splenocytes) are prepared on day 2 for immediate plating after blocking.

The spot development takes between 3 and 5 h, done on day 3.

Plates need to be dry before spots can be counted. It is hence recommended to run the plate evaluation on day 4 or later. Spots are stable for an extended time (months and even years). Plates should be stored protected from light. Even fluorescent spots are stable for some time. Bleaching of spots is dye dependent and caused by exposure to a strong light source as required for the automated Fluorospot evaluation. Most sensitive to bleaching are FITC spots, while Cy3 and Cy5 are much more stable. With optimized, commercially available Fluorospot kits, spot stability can reach weeks and months, if plates are stored light protected. Fluorescence enhancer may resurrect loss of spot intensity.

Chapter 6
The Elispot Assay

As an important resource, a step-by-step Elispot protocol with important annotations is provided elsewhere [79].

6.1 Coating

Coating is typically done the day before the assay, and the plate is incubated overnight at 4 °C. Details and logistics about the coating concentration are already provided in Sect. 3.2. Please review Fig. 3.5 about the influence of the coating concentration on spot size and staining intensity. If spots appear overall faint and rather diffused, it is likely that an increase in the coating concentration for the capture antibody (or antigen in case of B-cell Elispot) will improve this issue and lead to smaller, denser, better defined spots. It is recommended to follow the manufacturer's instructions on the recommended coating concentration when using commercial Elispot kits. For most applications it is sufficient to coat the plate with 100 µL of coating buffer (e.g., PBS without calcium or magnesium) containing 10–15 µg/mL capture antibody.

One very important issue is the necessary prewetting of PVDF ("IP") plates. Already in Sect. 3.1 we discussed the requirement to overcome the hydrophobicity of PVDF in order to ascertain optimal binding of the capture antibody or antigen, similar to prewetting the PVDF membrane in Western blot experiments. The following steps are essential for successful prewetting:

1. Only use absolute, non-denatured ethanol to prepare the stock solution. Denatured ethanol contains additives that may interfere with the membrane and overall assay outcome.
2. Use a freshly prepared wetting solution that contains between 35 and 70% ethanol. Some evidence exists that 35% ethanol leads to a slightly higher binding efficiency of antibody after prewetting. However, ethanol concentrations below

© Springer International Publishing Switzerland 2016
S. Janetzki, *Elispot for Rookies (and Experts Too)*, Techniques in Life Science and Biomedicine for the Non-Expert, DOI 10.1007/978-3-319-45295-1_6

Fig. 6.1 Well image with signs of impaired spot development due to leakage. Spots in the periphery are very faint and appear washed out due to direct contact of the membrane with ethanol that leaked into the underdrain.

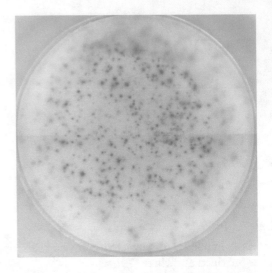

35% do not wet the membrane anymore; hence it is more reliable to use a 70% ethanol solution.

3. Only use 15 μL of ethanol to prewet. This is extremely important. The volume is sufficient to wet the membrane surface. Higher volumes (like 50 μL or higher) will lead to leakage of ethanol into the underdrain that is attached to the plate bottom. Ethanol will remain in the underdrain and promote leakage throughout the experiment. Spots may appear "washed out," and extremely faint, or not be visible at all (Fig. 6.1). Tap plate to ensure even distribution of this small amount of ethanol across the entire well. The membrane will turn dark grey-blue, a sign that it has been successfully prewet.

4. Do not incubate the plate with ethanol for a prolonged time. Wells may dry out and return to the high-hydrophobicity state they were at before adding the ethanol solution. Instead wash the plate three times with PBS. Leave the last wash of PBS in the plate until you have prepared the coating solution. Then dump the PBS and add the coating solution.

5. Do NOT prewet nitrocellulose plates.

Another important topic related to coating is pre-coating, as already reviewed in Sect. 1.3.1.1. Pre-coating plates in advance is attractive since plates are available for immediate use. Commercially pre-coated plates are available, which contain a stabilized coating antibody. However, do not pre-coat plates yourself, since the lack of stabilizers will lead to inconsistent binding of the capture antibody to the membrane, as illustrated in Fig. 1.5.

6.2   Blocking

After coating the plate, remove unbound antibody (or antigen) by washing the plate 1–2 times (using 150–200 μL blocking buffer per well is efficient). More washes are not necessary. Then add between 100 and 200 μL blocking buffer (as reviewed in

Sect. 3.6) per well and incubate the plate for at least 2 h at 37 °C (incubator). This step is important to block any nonspecific binding sites on the membrane (remember: membranes used in Elispot have a very high protein-binding capacity, which is only used at a low percentage when coating).

6.3 Effector Cell Preparation

The blocking time can be efficiently used to prepare the effector cells for plating. By now you have either obtained:

(a) Fresh cells

- If PBMC, then optimally rested overnight to reset cells to a tissue-like state as reviewed earlier (see Sect. 4.7);
- Or cells obtained from tissue, most likely spleen cells in case of working in the murine system, optimally with red blood cells removed by ACK lysis (see Sect. 4.3 and [21]);

(b) Frozen cells, thawed and PBMC optimally rested overnight (see Sects. 4.4, 4.5, and 4.7);

(c) Fresh or frozen cells, further manipulated:

- Isolated subpopulations of cells (e.g., CD8+ T cells; see Sect. 4.8),
- Short-term stimulated cells (e.g., for GrB release measurement [80], or after R848/IL-2 stimulation for antibody secretion [6]; both designed to convert cells from the memory to the activated effector state),
- In vitro-expanded cells (see Sect. 4.9).

Two important steps are necessary: (1) efficient washing and (2) accurate counting of cells.

Cells need to be washed to remove any impurities that would affect the assay, including

(a) Ficoll, if fresh PBMC were just isolated (and not further rested),
(b) Cell debris and impurities from lysed cells that underwent apoptosis (especially after overnight resting),
(c) Benzonase or other DNase products used to prevent clumping when thawing cells (see Sect. 4.5),
(d) ACK buffer and red blood cell debris after lysis when using spleen cells,
(e) Buffers and magnetic beads removed (if applicable) from cells after isolation of cell subpopulations,
(f) Cytokine or antibodies that have been released during short- or long-term culture (activation or expansion) of cells: This washing step is very important since existing analyte in the medium will bind to the capture antibody and cause high background staining.

Cells should be washed with the culture medium used for the assay (reviewed in Sect. 3.5). Please note that each washing step may promote some cells loss; hence only perform the absolutely necessary number of washing steps (in most cases two washes are sufficient).

Cell counting has been reviewed extensively in Sect. 4.6. As elaborated on earlier, to prepare the samples with the correct amount of living cells, it is advised to use a counting method that distinguishes dead and apoptotic cells from living cells. Usually, a small sample of the cells is obtained before the last washing step, so that cells can be resuspended in the appropriate volume after washing, ready to be plated out (see Sect. 6.6). In the same chapter we will review necessary cell concentrations which depend on the overall assay setup (stimulation).

6.4   Stimulant Preparation

Various forms of stimulants exist that cells can be exposed to, to check on specific and nonspecific analyte secretion.

6.4.1   Peptides and Peptide Pools

Peptides and peptide pools belong to the most common antigens used for stimulation in Elispot. Peptides are short motifs of amino acids. They differ in their length, with shorter peptides (typically consisting of 8–11 amino acids = 8–11 mers) binding to MHC class I molecules (the binding to a specific allele depends on their amino acid sequence), and longer peptides (about 15 amino acids long) preferentially binding to MHC class II molecules [81, 82]. While there exist many natural occurrences of peptides, in the context of this book it is sufficient to know that peptides are typically derived from the intracellular degradation of proteins and their transport to the cell surface where they are presented by MHC molecules. The selection of peptides presented on a cell is dependent on the proteins expressed by the cell or internalized (for example proteins from a virus that infected the cell [83]), and its MHC = HLA (human leucocyte antigen in humans) repertoire. A person expresses a variety of different HLA types (alleles), with some being more commonly expressed in certain subpopulations, e.g., HLA-A2.1.

Countless peptides have been identified and mapped for their origin (protein) and their MHC binding selection. This has high relevance for the vaccine development in the infectious disease and cancer field. T-cell responses are directed against peptides presented in MHC-restricted matter by cells. We can hence investigate the existence of antigen-specific responses by using peptides in Elispot.

Peptides can be directly added to PBMC or splc. If the MHC they specifically bind to is expressed by cells in that sample, they can bind directly to empty molecules on the cell surface [84, 85]. Hence, peptides added to PBMC can be presented by MHC molecules on the surface of B cells, T cells, and monocytes, It is therefore not necessary (though often suggested in older literature) to pulse other antigen-presenting cells

(APC) with peptide and then add them to the effector cells in the plate for stimulation if whole PBMC or splc preparations are used in the Elispot assay.

Longer peptides are assumed to be clipped by exogenous proteases or are taken up by APC for further processing (digestion) and loaded onto MHC molecules transported to the cell surface [86, 87].

An important enrichment to the use of peptides was the development of overlapping peptide pools that cover the entire length of a protein. The standard format of such pools consists of 15 mers overlapping by 11 amino acids which allows the simultaneous assessment of CD8 and CD4 responses and covers all possible epitopes contained [44, 88]. Two important advantages come to mind:

1. Overlapping peptide pools do not require the identification or knowledge of specific peptides eliciting an immune response,
2. Overlapping peptide pools do not require knowledge about the HLA type of a donor or patient.

While it has been shown that an increasing number of peptides added to one well increases the likelihood for competition among peptides for binding sites to MHC molecules and hence may lead to lower spot numbers compared to the sum of all responses if peptides were tested alone [89], peptide pools containing as many as 138 peptides have been successfully used such that they are widely employed as control peptide pools (also see Sect. 5.1.2) [20, 44]. Smaller pools, consisting of 7–30 peptides, have recently been demonstrated to expose high sensitivity in effectively detecting single-peptide responses [90]. However, a recent publication suggested that 15 mers may not always lead to efficient CD8-restricted T cell response detection [90].

The purity of peptides used in an Elispot assay requires attention. The lower the purity grade, the higher is the likelihood of false-positive responses [91]. Peptide preparations can be contaminated with impurities like deletion peptides (novel peptides due to the omission of an amino acid during synthesis) or cross-contaminated with other peptides synthesized at another time, when appropriate cleaning of the synthesis system was skipped [92]. The latter possibility should be kept in mind if intending to use an unusually cheap peptide provider. Residual solvents used during the synthesis and purification can also contaminate peptide preparation and potentially inhibit T-cell responses, if peptides were not obtained freeze dried.

Lastly, peptides are typically dissolved in DMSO. However, DMSO is an oxidant, and oxidizes cysteine among some other amino acids. It is therefore strongly recommended to obtain freeze-dried aliquots of peptides and only dissolve aliquots on a need-to-use basis. Repeated thaw and freeze cycles of dissolved peptides should be avoided.

The overall recommendations is to obtain peptides and peptide pools from a reliable source that can provide documentation about synthesis and content of the peptide/peptide pool preparation, and store the peptides freeze dried at −20 °C, or, if dissolved in DMSO, at −80 °C. Use peptides with purity >70%, preferably >90%.

For optimal stimulation in an Elispot assay it may be required to work out the optimal peptide concentration in titration experiments, during which cells are stimulated with different concentrations of peptides. An important limiting factor (next to the costs of obtaining peptides, or course) is the final amount of DMSO, which must not exceed 1% due to its cell toxicity. As a rule of thumb, a final concentration of 1 µg/mL in Elispot works well for peptides with strong binding affinity to the MHC molecule. For example, the CEF peptide pool which contains defined 8–11 mers from CMV, EBV, and flu virus as stated earlier is commonly used at final concentrations of 1–2 µg/mL. Reports exist that responses may be inhibited if too high peptide concentrations are used [93].

Note that a given peptide amount by the manufacturer for a peptide pool refers to the amount of each peptide in that pool; for example, if you obtain 25 µg of the CMVpp65 peptide pool, you will obtain 25 µg of each of the 138 peptides contained in that pool. The same applies when concentrations are used (e.g., 1 µg/mL of CEF peptide pool means that 1 µg/mL of each peptide is being used).

For plating, peptide working solutions have to be prepared in double as high concentration as the wanted final concentration due to the 1:1 dilution with cells (e.g., if the desired final concentration for stimulation is 5 µg/mL, a solution containing 10 µg/mL peptide has to be prepared).

6.4.2   Proteins and Lysate

Entire proteins and protein-containing lysate (e.g., from virus-infected cells) can also be used for stimulation, but they require uptake and processing by APC [94]. Hence, the incubation time for the assay should be increased (typically by 1 day). It may be of advantage to pulse professional antigen presenters (dendritic cells = DCs), which can be prepared in advance. The protocol is complex and time consuming, and will not be reviewed here. Examples for the preparation of cancer cell lysate and dendritic cells can be found in abundance in published literature [95, 96].

6.4.3   Antigen Presenters

There are two general kinds of APC:

1. Cells that were loaded or pulsed with antigen
2. Cells that express antigen of interest naturally

We have already reviewed some choices for APC that can be loaded or pulsed with antigen. In the previous subchapter we listed the option to prepare autologous dendritic cells from monocytes and mature them with cell lysate or protein, accompanied with references for detailed protocols.

In the earlier peptide subchapter, we also discussed the option to use the plated PBMC or splc population as effectors and antigen presenters simultaneously when using peptides.

A similar approach can be used when working with isolated subpopulations of cells, e.g., CD8+ T cells. This is sometimes done when the sensitivity of the assay should be increased. CD8+ T cells are isolated, plated out, and then carefully over-laid with the autologous CD8– cell population (or syngeneic naïve splc) pulsed with peptides (typically done at 37 °C for one 1 h in serum-free medium).

A still rather frequently used, but by far less optimal solution is the use of other, HLA-matched APC, most commonly perhaps T2 cells, a TAP-deficient human lym-phoid cell line that synthesizes uncomplexed HLA-A2.1 MHC molecules. Background reactivity against that cell line is common [97]. While T2 cells are defective in peptide transport due to their TAP deficiency, some peptides are pre-sented TAP-independently [98, 99], giving a possible explanation for the frequently observed background reactivity. The expression of low levels of allogeneic MHC molecules may be a further cause for that observation [100]. Lastly, T2 cells were suggested to have a higher susceptibility to NK cell reactivity [101]. Similar obser-vations have been made for the C1R cell line [102]. And if this is not reason enough to avoid the use of allogenic antigen presenters, then it is important to note that one of the recommendations from the initial CIMT Elispot proficiency panel testing is the avoidance of allogenic antigen presenters due to the suboptimal results they provided, compared to the direct addition of peptides to plated PBMC [65].

A different story goes along with cells that naturally present antigens, namely autologous or syngeneic tumor cells. Such cells are rather easily available in the murine system. Autologous tumor sample availability is likely to be restricted in humans for the majority of cases, but if available they have been shown to elicit detectable responses in Elispot [103, 104].

The cell number and concentration that need to be prepared are dependent on the setup of the Elispot assay (see Sect. 6.6). Cells need to be washed and counted, simi-larly to the effector cells. It is not required to assess their apoptosis degree.

6.4.4   Control Stimulants

An extensive list of positive control stimulants for nonspecific stimulation of cells has been provided earlier in Sect. 5.1.2. These serve as an overall assay control (the assay worked) as well as a sample control (the sample is responsive). The most commonly used positive control stimulants in cytokine Elispot assays are PHA-L, ConA, and PMA/ionomycin. Final working concentrations are also given in the overview provided earlier. You need to prepare solutions for plating that is double the concentration as required in the final assay, owing to the 1:1 dilution with the effector cells.

The essential negative control, medium, should contain the same amount of DMSO as contained in the peptide/peptide pool preparations (e.g., if the DMSO concentration in a peptide pool that is being added to the effector cells is 0.2%, then the medium that is added to cells alone also needs to contain 0.2% DMSO) [101].

6.5   Reference Samples

Reference samples follow the same preparation rules as test samples, PBMC or splc. Their use as an external control has been reviewed in Sect. 5.1.3.

6.6   Plating Out Cells and Stimulants

Three important rules have to be followed when plating out cells. These rules are:

1. "Happy" conditions for effector cells must be provided.
2. All effector cells must "see" the antigen.
3. Effector cells and capture antibody must be in direct contact.

For T-cell Elispot, there are two main scenarios that apply. In the first scenario, the plated cells serve as effector and APC simultaneously, e.g., when testing PBMC or splc plus peptides. In such case, enough cells have to be plated so that efficient antigen presentation is ensured, typically between 200,000 and 400,000 cells per well. This leads to a slight pileup of cells. It is important to realize that cells in the second layer without immediate contact with the capture antibody will not produce spots. However, the beginning pileup of cells provides optimal conditions for antigen presentation and co-stimulation (cells are "happy") [105]. The scenario is illustrated in Fig. 6.2.

The restriction of this scenario is that it delivers only a limited range of linear behavior, meaning that a titration of cells added per well does not correlate with the

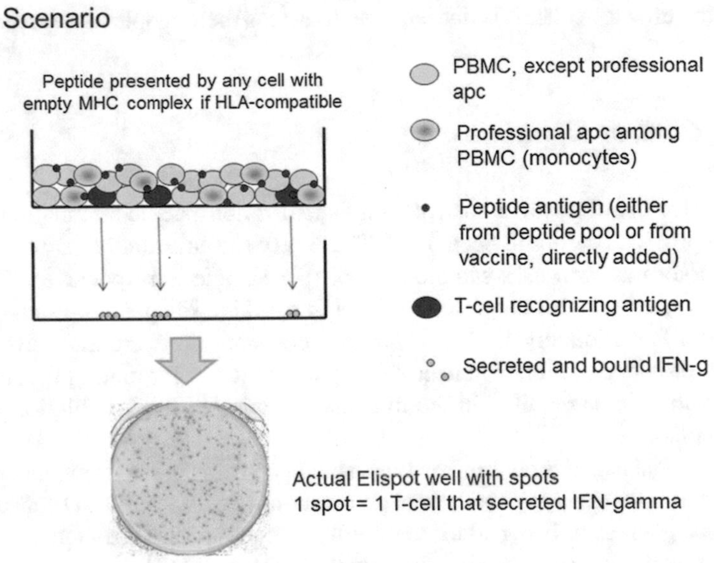

Fig. 6.2   Elispot scenario for plating cells and peptide antigen.

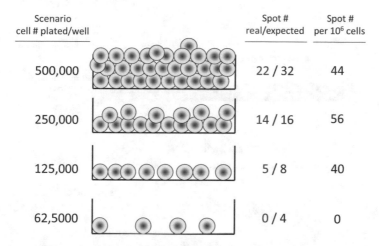

Scenario cell # plated/well		Spot # real/expected	Spot # per 10⁶ cells
500,000		22 / 32	44
250,000		14 / 16	56
125,000		5 / 8	40
62,5000		0 / 4	0

Fig. 6.3 Nonlinear decrease of the spot number in PBMC titration using peptide as antigen. Cells (*yellow–orange circles*) are plated in a well at different cell concentrations. The best conditions for the detection of antigen-specific T cells ("happy," "see" antigen, contact to capture antibody) are given with slight pileup of cells, here at 250,000 cells per well. Lower cell numbers per well (125,000 or 62,500) do not provide two of the three essential prerequisites for cytokine secretion for all cells (no "antigen," not "happy"); hence spot numbers decrease disproportionally fast. When adding too many cells per well (500,000), cells pile up in multiple layers, and while cells see the antigen and receive efficient co-stimulation, the third requirement, the direct contact with the cap-ture antibody, is not given for all cells; hence the spot number increase is disproportionally low. An eye-opening step to get a grasp on the actual detected frequency of antigen-specific T-cell is to extrapolate the spot number per condition to one million cells (e.g., multiplying the spot number of 22 obtained with 500,000 cells by 2=44 spots, or similarly multiplying the spot number of 14 obtained with 250,000 cells by 4=56 spots).

change in spot number. This has substantial consequences when cells are cut below a cell number that follows the rules listed above. When cells do not touch each other anymore, they are not able to "see" the antigen, and they are not "happy" due to missing cell-to-cell contact. On the other hand, increasing the cell number to a point where cells pile up in multiple layers does not lead to an according increase in the resulting spot number, due to the lack of contact between cells in the upper layer and the capture antibody on the membrane. The logistics of this issue are demonstrated in Fig. 6.3.

B cells do not entirely follow the same logistics, and are commonly titrated in Elispot experiments, with spot counts following a more linear behavior when cells do not touch each other anymore. But again, when too many cells are plated, cells that are not in direct contact with the capture antibody or antigen used for coating will secrete antibody into the culture medium, and cause high overall background staining, but do not produce spots.

The second plating scenario follows entirely different logistics. Here separate effectors and antigen presenters are used. It is important to avoid adding more cells than needed to form a monolayer of effector cells across the well. Cells should be let to settle down after plating before carefully overlaying them with the antigen

Scenario

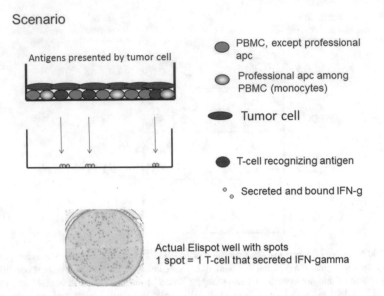

Antigens presented by tumor cell

- PBMC, except professional apc
- Professional apc among PBMC (monocytes)
- **Tumor cell**

- T-cell recognizing antigen

- $^{\circ}_{\,\circ}$ Secreted and bound IFN-g

Actual Elispot well with spots
1 spot = 1 T-cell that secreted IFN-gamma

Fig. 6.4 Elispot scenario for plating separate effector and antigen-presenting cells.

presenters. Key to this scenario is to use enough antigen presenters so that every effector cell below them can "see" the antigen. Effector cells can be titrated down, or used at numbers below the required amount for forming a monolayer, as long as the number of APC is kept high enough that, independent of "where" the effector cells are located in the well, they "see" the antigen. To ensure efficient co-stimulation, professional antigen presenters or other APC that are able to provide co-stimulatory signals are preferred for this setup. Most importantly, this setup does NOT follow any effector:target ratio rules, meaning when titrating the effector cell population, the target cell population needs to remain at the same concentration as used for higher effector cell numbers. This scenario is demonstrated in Fig. 6.4. Mixing of effector and APC should be avoided. First plate effectors, and then overlay them with APC.

The plating scenario determines the cell number that needs to be plated per well. For the first scenario, about 2×10^5–3×10^5 cells per well are optimally required. Typically cells are plated in 50 µL volume per well, and stimulator/controls are plated in 50 µL volume per well for a total of 100 µL per well. Hence, PBMC or splc should be prepared at concentrations of 4–6×10^6 cells per mL. For the second scenario, cells should be plated at 1–1.5×10^5 cells per well (or less, as discussed above), requiring cell solutions with a concentration of 2–3×10^6 cells/mL.

Add cells and antigen slowly to the wells to avoid uneven distribution of cells or, in the case of overlaying effector cells with APC or adding other stimulators to already plated effectors, to avoid disturbing the settled cell layer(s) (also see Section 3.8.1 and Fig. 3.8).

6.7   Incubation

Add the cover to the plates and carefully transfer them into an incubator with 37 °C, 5% CO_2. Ensure that racks within the incubator are leveled. Avoid repeated opening or closing of the incubator door, or other disturbances like vibration due to close-by centrifuges, all of which can cause cells to move and produce irregular spot patterns which may present challenges for spot counting (Fig. 6.5).

The necessary incubation time depends on the speed cytokine will be secreted, and has already been reviewed in Sect. 5.3.

6.8   Removal of Cells

Once the incubation has ended, it is not required to work under sterile conditions anymore. An important step now is the removal of all cells from the membrane before spots can be developed. We already reviewed in Sect. 3.1 the membrane features and their fine network of fibers (also see Fig. 3.1) to which cells stick to rather tightly. Hence efficient washing is crucial here. Solid pressure and high washing buffer volume, combined with a washing buffer using a suitable detergent, have been commonly and successfully used for this washing step. Repeated washing (six times) with PBS plus 0.05% Tween20 (also see Sect. 3.7) using a squirt bottle is simple and efficient. For high buffer flow and easy squeeze, consider cutting off the tip of the squirt bottle spout (see Fig. 6.6).

Automated washers can also be used, but do not possess any substantial advantage over washing by hand.

Fig. 6.5 Elispot well image with strongly irregularly shaped spots due to vibration. The image was taken with a Zeiss KS Elispot reader (Thornwood, NY).

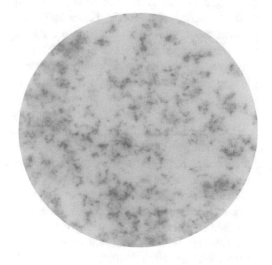

Fig. 6.6 Squeeze bottle with cutoff spout tip for washing plates after incubation.

6.9   Spot Development

The spot development for an enzymatic Elispot assay typically consists of three steps:

1. Adding a biotinylated detection antibody (or biotinylated antigen as discussed for B-cell Elispot in Sect. 1.3.2.1),
2. Adding the avidin-enzyme complex,
3. Adding the substrate.

For preparing the secondary antibody, manufacturers' recommendations should be followed closely. Overall, less amount of antibody is required compared to coating. However, aggregates of secondary antibody can cause false-positive spots (see Fig. 5.1). The amount of false-positive spots depends on the length of storage of the biotinylated antibody. Fresh antibody may only produce occasional false-positive spots, while antibody stored for many months or even years can produce false-positive spot numbers as high as 30 or even more per well. Hence it is imperative to remove any aggregates before plating the antibody. This can be achieved by filtering the secondary antibody solution after dilution with the appropriate buffer, just before plating with a low-protein-binding syringe filter, pore size 0.22 μm. The retention volume of these filters is neglectably small, and non-aggregated antibody easily passes through the filter, while aggregates are efficiently removed and with that the risk for any false-positive spots in the assay. Filtering the secondary antibody should be performed in any Elispot assay, independent of the antibody or manufacturer of the reagent kit.

The different choices for enzyme conjugates and related substrates have already been described in detail in Sect. 3.3. As a reminder, HRP in combination with TMB substrate has an outstanding sensitivity.

Incubation with the antibody should in average last about 2 h at 37 °C, followed by three washes with PBS/0.05% Tween 20. The incubation with the avidin–enzyme complex requires about 1 h (commonly done at room temperature).

Fig. 6.7 Pliers for
underdrain removal.

The detergent (Tween20) may interfere with the final development step; hence
it is recommended to wash the plate three times with PBS before adding the
substrate.

The incubation time with substrate depends on the substrate itself and the fresh-
ness (activity) of the enzyme. Spots normally appear within a few minutes, but may
not be recognizable immediately. AEC spots typically appear within 4 and 6 min,
and TMB spots between 4 and 10 min. BCIP spot development may vary more in
length. Stop spot development as soon as you see spots appear (spot development
can also be followed more closely under a dissecting microscope). Wash plates
under running water. TMB is sensitive to the ion content of water, and may produce
spot colors from dark blue (water ok) to light faint blue to red (water not optimal).
In the latter case, use deionized or distilled water for washing. Remove the underdrain from the back of the plate after spot development has been stopped. For HTS
plates, use pliers for underdrain removal (Fig. 6.7). Also wash the backside of the
membranes, remove remaining water by tapping the plate onto paper towels, and
use paper towels to carefully absorb remaining water/substrate from the backside of
the membranes.

Let plates dry completely before spot enumeration. Spots are stable for a long
time, typically months or even years in case of enzymatic Elispot (precipitated
substrate).

Fluorospot development does not use an enzyme–substrate sequential spot development, but employs fluorescent dyes instead. Follow the manufacturer's instructions closely. Use spot enhancer solution if provided, which boosts the spot intensity.
The incubation with spot enhancer takes about 15 min. Afterwards the enhancer
solution needs to be sufficiently removed by tapping the plate onto paper towels.
Fluorospots are stable for a prolonged time, at least days to weeks, if the plate is
stored in the dark.

6.10 Enumeration of Spots

The analysis of Elispot plates is a crucial step of an Elispot experiment. Already
during the early days of proficiency panel testing it was identified as a major con-
tributor to variability in results reported [20]. At first thought it may come as a surprise that nowadays the counting of colored spots on a white membrane is difficult,

considering the advances in imaging and computing. We will therefore review in more detail the challenges associated with the Elispot plate evaluation.

Excellent and detailed instructions on Elispot plate analysis including step-by-step guidance for any of the challenges reviewed below have been made available as guidelines obtained from a large international Elispot plate reading panel and consensus process with more than 100 scientists [19]. The publication also provides a flowchart that can easily be integrated in a laboratory's SOP concerning plate analysis, independent of the automated reader type used.

A comprehensive resource for Fluorospot plate analysis is also available [8].

The goal of Elispot evaluation is to obtain accurate counts of true spots in a well while attempting to exclude nonspecific background signals and any artificial signals. As a reminder, the hallmark of true spots is a dark center with fading color toward the spot periphery (also see Fig. 1.4). A spot is defined by

1. Its size (it needs to be larger than the cell that secreted the analyte due to its diffusion kinetics);
2. Its staining intensity (the more analyte bound across a certain area, the higher the staining intensity of a spot);
3. Its color (which is defined by the substrate used, also see Table 3.2);
4. Its shape (determined by the diffusion of the analyte away from the cell in all directions, therefore spots are more or less round);
5. Its staining gradient (dark center and fading staining toward the spot periphery, which is once again caused by the diffusion kinetics of the analyte).

An automated reader system should be able to distinguish spots based on the above features, and evenly importantly distinguish artificial signals and exclude them from being counted. This is often easier said than done, as demonstrated in Fig. 6.8.

The main sources of plate evaluation challenges can be summarized as follows:

1. The sample itself and everything that is contained in it,
2. Artificial disturbances not sample related,
3. Nonspecific background reactivity.

The sample itself can produce small little speckles caused by dying cells (Fig. 6.9). They often accumulate in the well periphery. Occurrence of such speckle rings is indicative of problems associated with the sample. They are the most common artifacts found in Elispot.

Other challenges caused by the sample are spot confluence (see Fig. 6.8b), tissue remains (see Fig. 6.8c), spot disintegration due to granulocyte contamination (see Fig. 4.2) or insufficient removal of cells from the membrane after incubation (Fig. 6.10).

Since the Elispot experiment is performed in an open environment, it can be subject to external interference, causing, for instance, pipette tip impressions on the membrane, accidental cracks of the membrane, or accumulation of dirt particles on the membrane. Spots can be irregularly shaped due to movement caused by vibration or rigorous handling of the plate or incubator (Fig. 6.5 and 6.8b), or they are too faint, e.g., due to insufficient coating (see Fig. 1.5) or leakage (see Fig. 6.1).

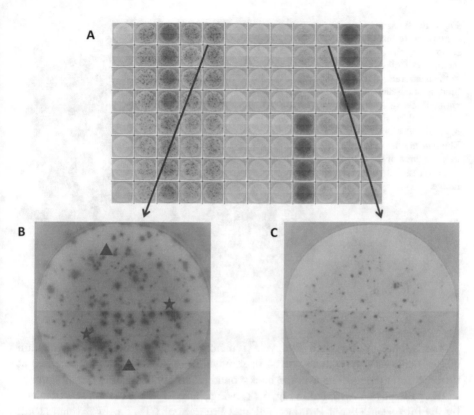

Fig. 6.8 Example of plate evaluation challenges. An overview of the entire plate is given (**a**). The plate was developed with BCIP/NBT for blue-purple spots. Wells appear to have different amounts of spots; from none visible (e.g., column 1) to many (e.g., columns 2–5), to very high numbers (e.g., column 9, four lower rows, and column 11, upper 4 rows). Some wells appear to have some spots (e.g., column 10). A closer look at well A5 (first row, 5th column) (**b**) reveals spots of different size, some of which appear confluent (*red stars*) and of odd shape (*red triangles*), likely due to movement of cells during incubation. Well A10 (first row, 10th column) (**c**) has multiple blue signals, but most, if not all, of them are artifacts. The main difference between signals in (**b**) and (**c**) is the fading of the staining from spot center to the periphery (**b**) or the lack thereof (**c**).

Fig. 6.9 Small artifacts in the well periphery (*blue arrow*). Spots are usually larger (as seen further toward the center of the well) and can thus be distinguished by appropriate size (and possibly intensity) adjustments of the reading parameters.

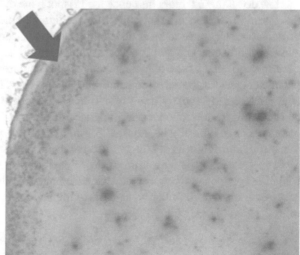

Fig. 6.10 *White spot* centers due to insufficient removal of cells after incubation. Cells sticking to the membrane block the binding of the detection antibody, causing white centers within otherwise regularly developed spots. The *blue arrows* randomly indicate three of the many examples in this well image.

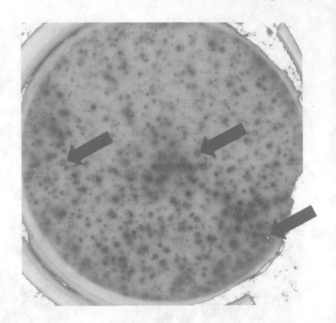

Lastly, background reactivity levels (typically assessed with the cells + medium control) can vary dramatically even in samples from the same donor, obtained at different time points. The average background reactivity of healthy donor samples in IFNγ Elispot is illustrated in Fig. 6.11, using a large number of data sets obtained by the European Elispot Proficiency Panel Program of CIMT over 3 years [106]. The graph summarizes the observed background reactivity reported by all participants, using different SOPs to test multiple healthy PBMC samples. It can hence be assumed with confidence that "normal" background reactivity levels in IFNγ Elispot are below 7 spots per 100,000 PBMC plated. This can of course vary between donors, and may be changed by treatment or infection status. Further, addition of cytokines to the Elispot assay or in vitro expansion of cells prior to the assay can increase background reactivity levels.

How do all these factors influence the Elispot plate evaluation? Key to answering this question is to accept that one set of reading parameters does not fit all conditions. Parameters have to be adjusted to a specific Elispot SOP and resulting spots, and constantly checked and if necessary adjusted for each sample tested. Sporadic occurrence of any of the above challenges may require the adjustment of reading parameters just for that specific condition. The question is this: How are parameters adjusted? Already at early times of automated Elispot evaluation it was shown that plate reading results obtained are operator dependent [24]. Two different sets of parameters can typically be adjusted for automated Elispot evaluation. Please note that different machines may apply different names to these parameters.

Set 1 relates to the algorithm of how spots are separated (or not). This is not trivial. Large spots are often somewhat inhomogeneously stained, and the software may recognize various darker centers within a spot. The separation algorithm

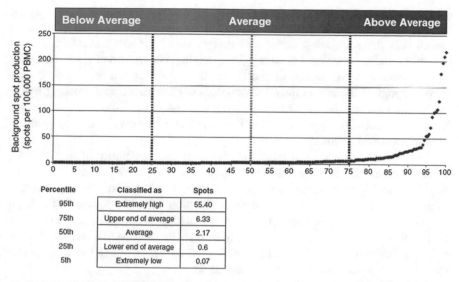

Percentile	Classified as	Spots
95th	Extremely high	55.40
75th	Upper end of average	6.33
50th	Average	2.17
25th	Lower end of average	0.6
5th	Extremely low	0.07

Fig. 6.11 Background spot production in IFNγ Elispot. 239 replicate data sets from three phases of Elispot proficiency panel testing by CIMT were plotted in ascending order. The x-axis shows the percentile rank and the y-axis the reported mean spot number for PBMC+ medium. Important percentile ranks are listed in the table below the graph. Background reactivity levels between the 25th and 75th percentile can be considered average. The figure was obtained from Moodie et al. [106].

determines if the spot is separated into multiple smaller spots which are then all counted individually, or if the large spot is only counted as one spot. Further, this algorithm determines how well spots are distinguished and counted when many spots are present in one well, elevating the overall background staining (but not necessarily consistently across the well) and also touching each other on occasion (though the spot centers are still recognizable). The sensitivity of spot separation clearly has to be set to a high level to allow the software to distinguish all apparent spots. Note that the higher the sensitivity for spot separation is set, the smaller will be the spot size reported. An overly high sensitivity can lead to false high spot counts, and vice versa spot separation of too low sensitivity may give rise to false low spot counts.

Set 2 is related to the parameters that define a spot, and typically include at least the spot size and the spot intensity (or contrast). Most reader programs also allow defining the color (or hue) of the spot, the spot shape (or roundness), and the degree of color fading from spot center to the spot periphery. The better the optics and resolution of the Elispot reader, and the larger the choice of parameters that can be adjusted, the easier it is to exclude artifacts from being counted. Further, the reading program may allow manual removal of counts caused by sporadic artifacts.

The parameter sets have to be fine-tuned using a well with an average amount of spots (if available) and comparing the outcome with the counts of a negative control well, using the same settings. Positive control wells (e.g., PHA-L) may have an abundance of spots, and may hence require an adjustment of the spot separation (which would not apply for the other wells). It is important to understand and accept the fact that once you determine the reading parameters, the working hypothesis is that all wells that require evaluation with that parameter set will present themselves in similar conditions (related to background staining, occurrence of artifacts, closeness of spots, and appearance of spots). As already elaborated on earlier, it may happen though that one specific condition does not fall under your working hypothesis (e.g., your positive control wells). Such wells require a different parameter set (e.g., adjustment of the spot separation algorithm). Reading results need to be annotated with such adjustments.

You also need to audit the plate and check on the plausibility of results (e.g., triplicates from the same condition with spot counts of 22–29–280 are not plausible). Wells and spots counted by the software need to be checked. Perhaps the differences in spot counts were caused by a pipetting mistake, perhaps by a large artifact (e.g., crack of membrane), or perhaps leakage occurred in the two of the three wells with low counts. Necessary adjustments for that specific well or specific wells are required, if feasible and plausible, and the results have to be annotated about the observation made and the action taken.

The overall message is: always check on the applicability of your reading parameters for each sample and be not afraid of adjusting parameters if necessary. Always audit your plate and readings, and make all necessary annotations that show what you did, where you did it, and why you did it. Most importantly: If the working hypothesis (that is related to spot definition, background staining, and artifact occurrence, etc., see further above) holds up for all wells, do NOT change the established parameters for any of the wells belonging to the sample. Even if the working hypothesis does not hold up for all conditions, attempt to read the according wells with the same parameter set. Only change the parameters for such wells, if spots are not picked up correctly or artifacts are included in the counts. Use common sense. You need to annotate the results about any irregular observations and parameter changes.

Refrain from using software options that automatically adjust parameters for you. They are based on substantially restricted working hypotheses, and the risk of obtaining simply wrong data is high.

First steps in the right direction have been taken that allow the overall harmonization of Elispot plate reading approaches. Please use the step-by-step guidance in Nature Protocols [19].

Chapter 7
Data Analysis

An Elispot experiment should be plated out in a way that each condition is tested in replicates, optimally at least in triplicates, if cell numbers allow (also see Fig. 5.2). Results of the final data analysis grow in strength the more negative control replicates are included, optimally six replicates [107]. Before analyzing the data for positivity, it makes sense to look at the variability between replicates (also check on the plausibility as discussed in the previous chapter). It can be expected that there is some variability between replicates (imagine the odds to have cells mixed and plated so perfectly that you end up, for example, with exactly 60 antigen-specific cells in 200,000 cells added to the well, in three consecutive wells). While some variability is acceptable, the question is this: How much variability is acceptable? Let's take a look at some hypothetical data sets:

Triplicates/Set 1:	88–98–93

The variability among these three replicates appears acceptable.

Triplicates/Set 2:	0–350–22

The variability among these three replicates is obviously not acceptable. But what about the third data set:

Triplicates/Set 3:	88–53–111

To answer this question, the variability between these replicates needs to be determined. This can be done using the following equation: intra-replicate variability = variance / (median + 1) [106]. The variability for Set 3 is 6.39. But what is this value really good for? A useful tool has been provided by the same authors who have addressed the average PBMC background reactivity (reviewed in the previous Sect. 6.10) [106]. Over 700 triplicate data sets from participants in three Elispot

© Springer International Publishing Switzerland 2016
S. Janetzki, *Elispot for Rookies (and Experts Too)*, Techniques in Life Science and Biomedicine for the Non-Expert, DOI 10.1007/978-3-319-45295-1_7

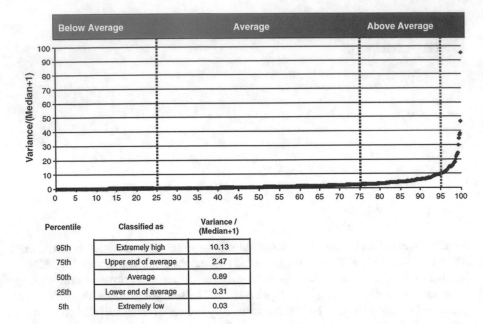

Fig. 7.1 Inter-replicate variability. 717 triplicate measurements from the CIMT Elispot panel were analyzed. Results were ascendingly ordered, with the percentile rank of the results shown on the *x*-axis, and the variability on the *y*-axis. Important percentile ranks are marked within the graph by a *dotted line*, and the corresponding variability values are listed in the table below the graph. The graph was obtained from Moodie, Price et al. [106].

proficiency panels (all using their own SOP, but testing the same PBMC in each panel) have been analyzed for their variability. The results are shown in Fig. 7.1.

The data indicate that variability <2.5 (using the equation given above) is average (or below average). Values above 10, on the other hand, are extremely high. Our third data set is above average, but not extremely high.

The variability equation can, for example, be integrated into Excel data sheets as a macro that flags replicates with extremely high (>10) variability.

The next question is how to define a response against an antigen, compared to the negative control. There are two different approaches possible:

1. An empirical approach, and
2. A statistical approach.

An empirical approach is an ad hoc tool, based on observations (= mean values) from a study. The following is an example of an empirical rule that defines a response when

(a) The mean of the experimental wells (e.g., PBMC + peptide pool) is at least 3× higher than the mean of the negative control (typically PBMC + medium); and
(b) The mean of the experimental wells is at least 10.

The challenge is that the set threshold does not take the inherent variability of Elispot data into account. This can lead to a lack in sensitivity and specificity. Here are two examples to explain these issues in more detail:

Data set/negative control: 6–8–10	Data set/experimental wells: 7–11–55

With the one large outlier value in the experimental data set this sample is defined as a positive responder (the mean of the experimental wells=24.3 which is >3× higher than the mean of the negative control=8, and is also larger than 10). This is likely a false-positive response call (plausibility!). Let's have a look at another data set:

Data set/negative control: 0–0–0	Data set/experimental wells: 9–7–11

Applying the above empirical rule, this sample is called negative because the mean of the experimental wells is below 10. It is easy, however, to agree that there is clearly "something."

To establish an empirical rule for a given study, multiple samples have to be tested that are not expected to have a response against the antigen of interest, in order to demonstrate that the established rule controls the false-positive rate. An excellent example for establishing an empirical rule is given elsewhere, and the reader is strongly advised to use this publication for guidance [108].

The requirements for a statistical tool are defined by the nature of Elispot data. Elispot data cannot be assumed to be normally distributed, simply because of the low amount of data points (typically triplicates). Normal distribution (Gaussian curve!) assumes that the mean equals the median that equals the mode (the data point most often represented in the set). Let's have a look at two data sets again:

Data set/negative control: 0–4–5	Data set/experimental wells: 4–56–6

The mean of the negative control (3) is similar to its median (4). But the mean of the experimental data set (22) is far off from the median (6). Statistical tests (for example, the Student's t-test) that assume a normal distribution would use the mean to determine the statistical significance of a response (p-value) here, and would likely report this sample to be a responder.

The solution to the dilemma is to use a nonparametric test that avoids assumptions about the data distribution. A well-suited nonparametric test for analyzing Elispot data for responses is the distribution-free resampling (DFR) method [106, 107, 109]. DFR testing takes the variability of Elispot data into account and controls the false-positive rate while maintaining sensitivity to each of the tested antigens. A free online tool is available: http://www.scharp.org/zoe/runDFR/.

To be able to use the online tool, Elispot data need to be structured and uploaded as a .csv data file as illustrated online. The instructions are easy to follow, and it takes only a few seconds to obtain the results. Two slightly different DFR tests will be run simultaneously, one more lenient test reporting any significant differences between experimental test wells and the negative control = DFR(eq), and one that is more strict and requires the experimental test data to be at least 2× higher than the negative control = DFR(2×).

The prerequisites for DFR testing are that testing was done at least in triplicates for each condition, and that at least two experimental conditions were tested in addition to the negative control.

An example for the results of DFR testing is given in Table 7.1:

The replicate values for the different test antigens are listed under well 1,2,3. The negative control values (6 replicates) are shown under c1–6. All four antigens are compared to the same negative control. The p-values for the DFR(eq) and DFR(2x) tests are given. Responses as established by each test are defined by either a "1" = positive response or a "0" = negative response.

To make response calls, we finally need to have a look at the limit of detection (LOD) of an Elispot assay. The LOD in Elispot defines the lowest spot number in a well that appears to be above background reactivity. It determines the grey area below which a quantitation of a response is not possible (meaning it cannot be reliably distinguished from background). In other words, even if DFR testing would identify statistical significant differences between the negative control and the experimental wells, the results should be flagged as being below the LOD and not be considered a response or marked as a questionable response that should trigger the repeat of the assay.

The LOD is assessed by determining the average background reactivity levels under a given SOP and using a signal:noise ratio of 2:1 (less stringent, when low background reactivity levels and low variability between different donors are observed) or 3:1 (more stringent, higher background reactivity levels and/or higher variability between different donors observed). If variability between replicates is very low, the standard deviation (SD) can be used instead of the signal:noise ratio. As an example, if the average background reactivity with a given SOP is three spots per well, then under the less stringent condition the LOD would be 6, and under the more stringent condition it would be 9.

Another response definition addresses the response to a specific treatment and compares spot numbers between time points (e.g., before and after vaccination). Identifying immune responders using Elispot in the clinical context presents a number of confounding factors, such as specific aspects of the patient population and the structure of the data, that can make simple, single-metric rules about what constitutes a "response" difficult to apply broadly. This is especially true when dealing with both pre- and posttreatment samples, either of which could theoretically show some level of response (compared to the background control, as reviewed above). A promising approach was recently taken by Radleigh Santos and Clemencia Pinilla at the Torrey Pines Institute for Molecular Studies. Their approach takes in multiple metrics, including spot count difference from controls, DFR/mDFR p-values, and other statistical techniques applied directly to pre- and posttreatment results, and can offer a more unbiased approach to the determination of responder status [110]. Combining metrics in such a "fusion" approach leads to its own challenge, however; while single metrics, such as p-values, are easy to interpret, numerical combinations

Table 7.1 DFR testing results.

ID	Day	Antigen	Well 1	Well 2	Well 3	c1	c2	c3	c4	c5	c6	DFR (eq) adjp	DFR (eq) response	DFR (2x) adjp	DFR (2x) response
113	7	Peptide 1	5	18	8	5	6	5	3	6	6	0.07143	0	0.858	0
113	7	Negpeptide	7	4	13	5	6	5	3	6	6	0.17857	0	0.858	0
113	7	Peptide 2	108	96	94	5	6	5	3	6	6	0.0119	1	0	1
113	7	CEF	380	404	381	5	6	5	3	6	6	0.0119	1	0	1

of such metrics can result in numerical scores without a clear context. Monte Carlo simulation using negative control data on a given fusion scoring system can put these numbers into context. In particular, critical values for fusion scores can be determined based on percentiles of the results of the simulation, thus identifying clear thresholds for Elispot immune response which take in all relevant metrics, are based on the underlying negative response of the given patient population, and preserve the data structure of the given trial. Further details are beyond the scope of this book. The interested reader is strongly advised to review the original publications [110, 111].

Chapter 8
Quality Control

Throughout the book we have touched upon various important controls, for the sample (apoptosis and viability, see Sect. 4.6), and for the assay (negative controls, sample-specific internal controls, nonspecific positive controls, external controls, see Sect. 5.1).

There are a few more important steps to take to ensure quality, especially when planning on applying Elispot for clinical sample testing. All of these steps are essential for providing confidence in the data obtained.

Three main challenges exist for tests like Elispot that functionally assess single immune cells. These challenges are the following:

1. The assays possess an inherent variability.
2. There is no true gold standard available to control the assay accuracy.
3. Many different SOPs exist for performing the same assay.

An early idea about the variability in immune monitoring was obtained during the first years of the proficiency panel program by CIC (Fig. 8.1).

For the Elispot proficiency panel the same samples and antigens were sent to laboratories that were experienced in Elispot and used the IFNγ assay for the assessment of clinical samples [20]. Laboratories were allowed to use their own SOP. The reported results were widely spread. Examples of well images from all 36 laboratories testing the same sample during the first CIC panel have been published, and powerfully highlight the variability issue [113]. The question arose who was actually right, or in other words, which results were indeed accurate. The challenge is that the accuracy could not be determined since no gold standard or gold standard test exist that can provide the exact answer about how many functionally active cells were contained in each PBMC sample. The only way to address accuracy in Elispot testing is to obtain the relative accuracy in performance by participating in proficiency panels. Since the same samples are sent out to all panel participants, such exercises provide a measure of agreement of own results with the results reported by many other laboratories which were obtained with a variety of different SOPs. Figure 8.1 illustrates that there is a good accumulation of results around

© Springer International Publishing Switzerland 2016
S. Janetzki, *Elispot for Rookies (and Experts Too)*, Techniques in Life Science and Biomedicine for the Non-Expert, DOI 10.1007/978-3-319-45295-1_8

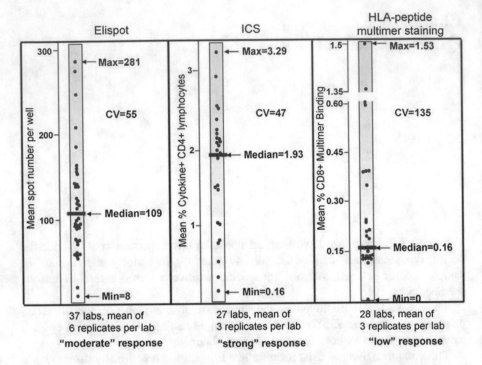

Fig. 8.1 Inter-laboratory variability in immune response measurement. The results of the first Elispot, intracellular cytokine staining (ICS), and HLA-peptide multimer staining panels of the CIC are shown. The *blue dots* represent the mean of reported replicate values for each participating laboratory. For each assay, a different PBMC sample was tested; however, all participants tested the same PBMC with the assay specified. The minimum and maximum reported values, the panel median, and coefficient of variation (CV) are indicated. The figure was obtained from the MIATA Reporting Guidelines paper in Immunity in 2009 [112].

the panel median. If the panel is large enough (many laboratories participate), it can be assumed that the panel median represents a good estimate of the true value. The relative accuracy describes the relation of the own measurement to that panel median. The closer the proximity of the relative accuracy to the value 1, the better the overall agreement of own results to results obtained from a representative cross-section of the field.

Example Set 1:

- Own measurement (mean of 3 replicates): 150
- Panel median: 142
- Relative accuracy: 1.06 (very good)

Example Set 2:

- Own measurement (mean of 3 replicates): 15
- Panel median: 142
- Relative accuracy: 0.1 (bad)

Initial Elispot Harmonization Guidelines to Optimize Assay Performance
A   Establish laboratory SOP for Elispot testing procedures, including:
A1   Counting method for apoptotic cells for determining adequate cell dilution for plating
A2   Overnight rest of cells prior to plating and incubation
B   Use only pre-tested and optimized serum allowing for low background : high signal ratio
C   Establish SOP for plate reading, including:
C1   Human auditing during reading process
C2   Adequate adjustments for technical artifacts
D   Only allow trained personnel, which is certified per laboratory SOP, to conduct assays

Fig. 8.2 Elispot harmonization guidelines. The figure was obtained from the original CIC panel paper [20].

There is currently one large proficiency panel program for Elispot available, and open to participation independent of the field a laboratory works in, its location, or affiliation. This program is conducted by Immudex. Detailed information and registration are available online at www.proficiencypanel.com.

The data obtained from the proficiency panels conducted by CIC and CIMT have also been used to identify critical protocol variables that influenced the overall performance of participating laboratories [20, 65]. Repeated rounds of panel testing provided proof for the validity of the initially identified variables, which were summarized as Elispot harmonization guidelines (Fig. 8.2) [20].

The increasing acceptance and integration of these harmonization guidelines correlate with lower variability and overall better Elispot performance of laboratories [64, 114]. When establishing Elispot in a laboratory, it is strongly recommended to harmonize the SOP with these guidelines. To harmonize an assay means to adapt critical protocol steps to optimize the assay outcome, based on guidelines obtained from iterative testing of the same samples with many SOPs (as done in proficiency panels).

Once the protocol has been optimized (and hopefully also standardized), it is important to demonstrate that the assay produces repeatable results with a low false-positive rate. This is done by assessing the precision of the assay. Other terms describing precision are repeatability or consistency. Three different precision measures should be evaluated:

- Intra-assay precision (between replicates) = repeatability
- Inter-assay precision (between assays) = intermediate repeatability
- Inter-operator precision = (between operators) = reproducibility

Precision testing involves testing of multiple samples (and multiple replicates per condition) on multiple days and by all operators that are involved in a study. A coefficient of variation (CV) below 50% should be reached, and a CV below 25% is desirable. However, CV values will rise dramatically, the lower the spot counts become. A high CV in low spot counts is a reflection of the mathematical issue related to the CV, which presents the ratio of the standard deviation (SD) to the mean. When the mean is very small or close to zero (= very low spot counts), the CV will approach infinity. Hence, the CV is not a usable tool to address variability in low spot counts. Here, the standard deviation should be used. The approximate spot count cutoff under which the CV dramatically increases lays around 30 spots [115].

Do not start a study if you do not know anything about the consistency in performance. Also, repetition of an experiment, if sample availability allows, is strongly recommended to deliver confidence in the results. The following example shall underline the importance of a tight precision and repetition of the experiment when feasible:

Sample 1 is tested for reactivity against medium = background control (BG) and a test antigen (TA) on 3 different days. The background counts were also subtracted from the test antigen counts (TA − BG). The mean spot counts are:
Sample 1:

	Day 1	Day 2	Day 3
BG:	5	2	8
TA:	12	6	20
TA − BG:	7	4	12

A different sample, Sample 2, is also tested for BG and TA reactivity on 3 different days. The mean spot counts are:

	Day 1	Day 2	Day 3
BG:	8	2	5
TA:	22	38	12
TA − BG:	14	36	7

Results for both samples over three experiments vary to a degree. The overall trend for each sample appears the same between all three testing dates. Is the precision acceptable? Let's take a pragmatic approach to answer this question:

Imagine that Samples 1 and 2 are samples from a clinical study, obtained from the same patient, Sample 1 before vaccination, and Sample 2 after vaccination. Your working hypothesis is that vaccination elicits or increases the immune response to TA. You test both samples in one experiment, and obtain results that match the day 1 precision testing results in our above examples:

	Before	After vaccination
BG:	5	8
TA:	12	22
TA − BG:	7	14

When comparing the background-corrected spot counts, it appears that the very weak preexisting response to TA about doubled after vaccination (from 7 to 14). To confirm this observation, the experiment is repeated. The spot counts obtained in the second experiment now match the day 2 precision testing results:

	Before	After vaccination
BG:	2	2
TA:	6	38
TA − BG:	4	36

The comparison of background-corrected spots reveals a 9-fold increase in the TA-specific response after vaccination (from 4 to 36 spots). This would clearly confirm the findings from the first experiment. But what if the repeat testing reveals spot counts that match the day 3 precision testing results?

	Before	After vaccination
BG:	8	5
TA:	20	12
TA − BG:	12	7

In this case there is no increase in response to TA upon vaccination (rather a decrease from 12 to 7 spots after background correction).

This example clearly demonstrates the importance of consistently and reliably performing the assay, and the necessity to show that you can perform an Elispot experiment with tight precision. Repeating an experiment, if anyhow feasible, is integral to providing confidence in your data.

Chapter 9
Reporting Guidelines

Have you ever tried to repeat somebody's experiment, but did not find the necessary information in the publication on how the experiment was conducted? Once you know about the critical protocol steps of an Elispot assay, don't you wonder how they were addressed by others? For example: How long it took to isolate PBMC after blood draw? How was resting performed, and for how long? What was the background reactivity for the tested samples? Was the serum used pretested? Were reported results background corrected? Or would you like to see an example image of crucial wells (which will likely be the best example available)? Did you wonder how the authors established reading parameters, and if reading parameters were adapted when necessary? Many more questions come to mind.

Transparency in reporting on assay conduct is a hot topic of discussion, and a multidiscipline-spanning project, the MIBBI (Minimum Information for Biological and Biomedical Investigations) project, was founded to provide an umbrella portal for reporting guidelines [116]. One of the early and most successful examples for such guidelines is MIAME which provides guidance on how to report on the conduct of microarray experiments [117, 118]. MIAME's usefulness in enabling others to understand and interpret data presented in publications led to its acceptance by scientists and impelled the guidelines to an obligatory publication standard by journals. Many other minimal information guidelines have been established since then, for example the MyFlowCyt guidelines for reporting on the conduct of flow cytometry experiments [119].

In 2009, the MIATA project was introduced to the field, addressing the minimal information about T-cell assays [112]. Spearheaded by a core team of scientists from various fields of immunology, the process of establishing these reporting guidelines involved two public consultation periods (online, open to everyone), and two public workshops with an invited panel of important stakeholders in immunology, followed by multiple webinars, to reach a consensus. A fine balance was reached between what scientists want and need to see in a publication related to how a T-cell assay was done (Elispot, ICS, multimer staining, and similar assays), and what scientists are willing to share in regard to the information requested and the

© Springer International Publishing Switzerland 2016 79
S. Janetzki, *Elispot for Rookies (and Experts Too)*, Techniques in Life Science and Biomedicine for the Non-Expert, DOI 10.1007/978-3-319-45295-1_9

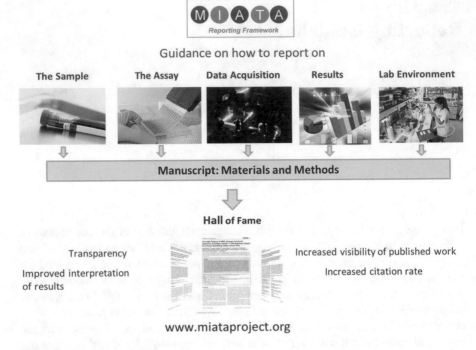

Fig. 9.1 Overview of MIATA.

effort to create MIATA-structured reports [120]. The final guidelines have been published [121], and a dedicated website was established (http://miataproject.org). The website provides guidance on how to structure the materials and methods section in a publication, and provides multiple tools like a checklist and pre-written sections for describing an Elispot, ICS, or multimer staining experiment in materials and methods sections of publications. These pre-written sections enable the author to fill in lab-specific information on a mouse click. MIATA-compliant publications are entered into an online Hall of Fame, which links to the authors' original publication (http://miataproject.org/implementation/hall-of-fame/). Not only does the MIATA Hall of Fame provide an easy tool to find publications with sufficient information on how experiments were conducted, but it also provides increased visibility to papers with transparent reporting on assay conduct. A summary of what MIATA is and what it offers is provided in Fig. 9.1.

Chapter 10
Troubleshooting

This book has been structured in a way that it provides detailed information on choices about how to run an optimized Elispot assay. Throughout the book examples are given for what works and what does not, for good and suboptimal outcomes; and recommendations on how to prevent suboptimal outcomes and the reasoning behind it are provided.

Therefore this section rather provides a summary (and reminder) of information that can be found in previous chapters.

No spots at all	Right antibody used?
	Right antibody concentration used?
	Expired reagents?
	Plates self-pre-coated and long-term stored?
	Think you used pre-coated plate, but used non-coated plate by mistake?
	Did you use reagents from a trusted vendor?
	Living cells added? Are you sure?
	Do the cells you added indeed secrete the analyte of interest?
Faint spots	Consider increasing coating concentration
	If PVDF plate used, did you prewet plate before coating?
	Check medium/serum (how old, serum pretested?)
	Expired reagents (check on enzyme, substrate, H_2O_2, if used)?
	Incubation time appropriate?
	Leakage due to too high prewetting volume?
Spot color changes across wells or plate	Use distilled or deionized water for stopping spot development and for washing plate after spot development was stopped
Spots are too large, but strong	Decrease incubation time of cells in plate
Spots are large, but faint	Increase coating concentration
Occasional blank wells	Prewet PVDF membrane
Partially blank wells	Prewet PVDF membrane

© Springer International Publishing Switzerland 2016
S. Janetzki, *Elispot for Rookies (and Experts Too)*, Techniques in Life Science and Biomedicine for the Non-Expert, DOI 10.1007/978-3-319-45295-1_10

Spots are accumulated on one side of well	Level incubator and incubator shelves
Spots are irregularly distributed (see Fig. 3.8)	Add cells slowly to well
	Check on pipet for smooth control over pipetting speed
	Carefully pipet antigen as not to disturb cell layer (or add before adding cells to well)
Irregular spot shape	Do not move plate during incubation
	Carefully open and close incubator while cells are incubating in Elispot plate
	Prevent vibration (e.g., centrifuge close to incubator)
Small artifacts (speckle-like)	Check on cell viability and apoptosis (small speckles are an indicator for cell death!!)
Large artifacts	Prevent DNA precipitates by using Benzonase®
	Filter cells obtained from tissue (e.g., spleen cells) with single-cell strainer
High variability between replicates	Use new HTS plates
	Use multichannel pipette and mix sample well
	Calibrate multichannel pipette
High background staining	Wash cells previously incubated (overnight resting, IVS)
	Shorten time for spot development
	Check on serum used (heteroclitic antibodies?)
Colored patch in well center	Remove all substrate and washing buffer after spot development
	Remove underdrain after spot development was stopped
High background reactivity	Pretest serum
	Did you filter the secondary antibody before plating to avoid false-positive spots?
	Reading parameters adjusted to exclude artifacts?

References

1. Czerkinsky CC, Svennerholm AM. Ganglioside GM1 enzyme-linked immunospot assay for simple identification of heat-labile enterotoxin-producing Escherichia coli. J Clin Microbiol. 1983;17(6):965–9.
2. Czerkinsky CC, Nilsson LA, Nygren H, Ouchterlony O, Tarkowski A. A solid-phase enzyme-linked immunospot (ELISPOT) assay for enumeration of specific antibody-secreting cells. J Immunol Methods. 1983;65(1-2):109–21. doi:10.1016/0022-1759(83)90308-3.
3. Czerkinsky C, Andersson G, Ekre HP, Nilsson LA, Klareskog L, Ouchterlony O. Reverse ELISPOT assay for clonal analysis of cytokine production. I. Enumeration of gamma-interferon-secreting cells. J Immunol Methods. 1988;110(1):29–36. doi:10.1016/0022-1759(88)90079-8.
4. Cox JH, Ferrari G, Janetzki S. Measurement of cytokine release at the single cell level using the ELISPOT assay. Methods. 2006;38(4):274–82. doi:10.1016/j.ymeth.2005.11.006.
5. Herr W, Linn B, Leister N, Wandel E, Meyer zum Büschenfelde KH, Wolfel T. The use of computer assisted video image analysis for the quantification of CD8+ T lymphocytes producing tumor necrosis factor alpha spots in response to peptide antigens. J Immunol Methods. 1997;203(2):141–52. doi: http://dx.doi.org/10.1016/S0022-1759(97)00019-7.
6. Jahnmatz M, Kesa G, Netterlid E, Buisman AM, Thorstensson R, Ahlborg N. Optimization of a human IgG B-cell ELISpot assay for the analysis of vaccine-induced B-cell responses. J Immunol Methods. 2013;391(1-2):50–9. doi:10.1016/j.jim.2013.02.009.
7. Boulet S, Ndongala ML, Peretz Y, Boisvert MP, Boulassel MR, Tremblay C, et al. A dual color ELISPOT method for the simultaneous detection of IL-2 and IFN-gamma HIV-specific immune responses. J Immunol Methods. 2007;320(1-2):18–29. doi:10.1016/j.jim.2006.11.010.
8. Janetzki S, Rueger M, Dillenbeck T. Stepping up ELISpot: multi-level analysis in FluoroSpot assays. Cells. 2014;3(4):1102–15. doi:10.3390/cells3041102.
9. Gazagne A, Claret E, Wijdenes J, Yssel H, Bousquet F, Levy E, et al. A Fluorospot assay to detect single T lymphocytes simultaneously producing multiple cytokines. J Immunol Methods. 2003;283(1-2):91–8. http://dx.doi.org/10.1016/j.jim.2003.08.013.
10. Ahlborg N, Axelsson B. Dual- and triple-color fluorospot. Methods Mol Biol. 2012;792:77–85. doi:10.1007/978-1-61779-325-7_6.
11. Dillenbeck T, Gelius E, Fohlstedt J, Ahlborg N. Triple cytokine FluoroSpot analysis of human antigen-specific IFN-gamma, IL-17A and IL-22 responses. Cells. 2014;3(4):1116–30. doi: 10.3390/cells3041116.
12. Körber N, Behrends U, Hapfelmeier A, Protzer U, Bauer T. Validation of an IFNgamma/IL2 FluoroSpot assay for clinical trial monitoring. J Transl Med. 2016;14(1):175. doi:10.1186/s12967-016-0932-7.

13. Kesa G, Larsson PH, Ahlborg N, Axelsson B. Comparison of ELISpot and FluoroSpot in the analysis of swine flu-specific IgG and IgA secretion by in vivo activated human B cells. Cells. 2012;1(2):27–34. doi:10.3390/cells1020027.

14. Jahnmatz P, Bengtsson T, Zuber B, Farnert A, Ahlborg N. An antigen-specific, four-color, B-cell FluoroSpot assay utilizing tagged antigens for detection. J Immunol Methods. 2016. doi:10.1016/j.jim.2016.02.020.

15. Hadjilaou A, Green AM, Coloma J, Harris E. Single-cell analysis of B cell/antibody cross-reactivity using a novel multicolor fluorospot assay. J Immunol. 2015;195(7):3490–6. doi:10.4049/jimmunol.1500918.

16. Smedman C, Ernemar T, Gudmundsdotter L, Gille-Johnson P, Somell A, Nihlmark K, et al. FluoroSpot analysis of TLR-activated monocytes reveals several distinct cytokine secreting subpopulations. Scand J Immunol. 2011. doi:10.1111/j.1365-3083.2011.02641.x.

17. Weiss AJ. Membranes and membrane plates used in ELISPOT. Methods Mol Biol. 2005;302:33–50. doi:10.1385/1-59259-903-6:033.

18. Weiss AJ. Overview of membranes and membrane plates used in research and diagnostic ELISPOT assays. Methods Mol Biol. 2012;792:243–56. doi:10.1007/978-1-61779-325-7_19.

19. Janetzki S, Price L, Schroeder H, Britten CM, Welters MJ, Hoos A. Guidelines for the automated evaluation of Elispot assays. Nat Protoc. 2015;10(7):1098–115. doi:10.1038/nprot.2015.068.

20. Janetzki S, Panageas KS, Ben-Porat L, Boyer J, Britten CM, Clay TM, et al. Results and harmonization guidelines from two large-scale international Elispot proficiency panels conducted by the Cancer Vaccine Consortium (CVC/SVI). Cancer Immunol Immunother. 2008;57(3):303–15. doi:10.1007/s00262-007-0380-6.

21. Janetzki S, Cox JH, Oden N, Ferrari G. Standardization and validation issues of the ELISPOT assay. Methods Mol Biol. 2005;302:51–86. doi:10.1385/1-59259-903-6:051.

22. Malyguine A, Strobl SL, Shafer-Weaver KA, Ulderich T, Troke A, Baseler M, et al. A modified human ELISPOT assay to detect specific responses to primary tumor cell targets. J Transl Med. 2004;2(1):9. doi:10.1186/1479-5876-2-9.

23. Janetzki S, Price L, Britten CM, van der Burg SH, Caterini J, Currier JR, et al. Performance of serum-supplemented and serum-free media in IFNgamma Elispot Assays for human T cells. Cancer Immunol Immunother. 2010;59(4):609–18. doi:10.1007/s00262-009-0788-2.

24. Janetzki S, Schaed S, Blachere NE, Ben-Porat L, Houghton AN, Panageas KS. Evaluation of Elispot assays: influence of method and operator on variability of results. J Immunol Methods. 2004;291(1-2):175–83. doi:10.1016/j.jim.2004.06.008.

25. Hoffmeister B, Bunde T, Rudawsky IM, Volk HD, Kern F. Detection of antigen-specific T cells by cytokine flow cytometry: the use of whole blood may underestimate frequencies. Eur J Immunol. 2003;33(12):3484–92. doi:10.1002/eji.200324223.

26. Mallone R, Mannering SI, Brooks-Worrell BM, Durinovic-Bello I, Cilio CM, Wong FS, et al. Isolation and preservation of peripheral blood mononuclear cells for analysis of islet antigen-reactive T cell responses: position statement of the T-Cell Workshop Committee of the Immunology of Diabetes Society. Clin Exp Immunol. 2011;163(1):33–49. doi:10.1111/j.1365-2249.2010.04272.x.

27. CLSI. Performance of single cell immune response assays; approved guideline – Second edition. CLSI document I/LA26-A2. Wayne, PA: Clinical and Laboratory Standards Institute; 2013.

28. CLSI. Procedures for the collection of diagnostic blood specimen by venipuncture; approved standard – sixth edition. CLSO document GP41-A6. Wayne, PA: Clinical and Laboratory Standards Institute; 2007.

29. Bull M, Lee D, Stucky J, Chiu YL, Rubin A, Horton H, et al. Defining blood processing parameters for optimal detection of cryopreserved antigen-specific responses for HIV vaccine trials. J Immunol Methods. 2007;322(1-2):57–69. doi:10.1016/j.jim.2007.02.003.

30. Thornthwaite JT, Rosenthal PK, Vazquez DA, Seckinger D. The effects of anticoagulant and temperature on the measurements of helper and suppressor cells. Diagn Immunol. 1984;2(3):167–74.

31. Boyum A. Isolation of mononuclear cells and granulocytes from human blood. Isolation of monuclear cells by one centrifugation, and of granulocytes by combining centrifugation and sedimentation at 1 g. Scand J Clin Lab Invest Suppl. 1968;97:77–89.

32. Afonso G, Scotto M, Renand A, Arvastsson J, Vassilieff D, Cilio CM, et al. Critical parameters in blood processing for T-cell assays: validation on ELISpot and tetramer platforms. J Immunol Methods. 2010;359(1-2):28–36. doi:10.1016/j.jim.2010.05.005.

33. Ruitenberg JJ, Mulder CB, Maino VC, Landay AL, Ghanekar SA. VACUTAINER CPT and Ficoll density gradient separation perform equivalently in maintaining the quality and function of PBMC from HIV seropositive blood samples. BMC Immunol. 2006;7:11. doi:10.1186/1471-2172-7-11.

34. Kierstead LS, Dubey S, Meyer B, Tobery TW, Mogg R, Fernandez VR, et al. Enhanced rates and magnitude of immune responses detected against an HIV vaccine: effect of using an optimized process for isolating PBMC. AIDS Res Hum Retroviruses. 2007;23(1):86–92. doi:10.1089/aid.2006.0129.

35. Kaplan J, Nolan D, Reed A. Altered lymphocyte markers and blastogenic responses associated with 24 hour delay in processing of blood samples. J Immunol Methods. 1982;50(2):187–91.

36. Schmielau J, Finn OJ. Activated granulocytes and granulocyte-derived hydrogen peroxide are the underlying mechanism of suppression of t-cell function in advanced cancer patients. Cancer Res. 2001;61(12):4756–60.

37. McKenna KC, Beatty KM, Vicetti Miguel R, Bilonick RA. Delayed processing of blood increases the frequency of activated CD11b+ CD15+ granulocytes which inhibit T cell function. J Immunol Methods. 2009;341(1-2):68–75. doi:10.1016/j.jim.2008.10.019.

38. De Rose R, Taylor EL, Law MG, van der Meide PH, Kent SJ. Granulocyte contamination dramatically inhibits spot formation in AIDS virus-specific ELISpot assays: analysis and strategies to ameliorate. J Immunol Methods. 2005;297(1-2):177–86. doi:10.1016/j.jim.2004.12.009.

39. Zea AH, Rodriguez PC, Culotta KS, Hernandez CP, DeSalvo J, Ochoa JB, et al. L-Arginine modulates CD3zeta expression and T cell function in activated human T lymphocytes. Cell Immunol. 2004;232(1-2):21–31. doi:10.1016/j.cellimm.2005.01.004.

40. Zea AH, Rodriguez PC, Atkins MB, Hernandez C, Signoretti S, Zabaleta J, et al. Arginase-producing myeloid suppressor cells in renal cell carcinoma patients: a mechanism of tumor evasion. Cancer Res. 2005;65(8):3044–8. doi:10.1158/0008-5472.CAN-04-4505.

41. Lenders LM, Meldau R, van Zyl-Smit RN, Woodburne V, Maredza A, Cashmore TJ, et al. Comparison of same day versus delayed enumeration of TB-specific T cell responses. J Infect. 2010;60(5):344–50. doi:10.1016/j.jinf.2010.01.012.

42. Bouwman JJ, Thijsen SF, Bossink AW. Improving the timeframe between blood collection and interferon gamma release assay using T-Cell Xtend((R)). J Infect. 2012;64(2):197–203. doi:10.1016/j.jinf.2011.10.017.

43. Currier JR, Kuta EG, Turk E, Earhart LB, Loomis-Price L, Janetzki S, et al. A panel of MHC class I restricted viral peptides for use as a quality control for vaccine trial ELISPOT assays. J Immunol Methods. 2002;260(1-2):157–72. doi:S002217590100535X [pii].

44. Maecker HT, Dunn HS, Suni MA, Khatamzas E, Pitcher CJ, Bunde T, et al. Use of overlapping peptide mixtures as antigens for cytokine flow cytometry. J Immunol Methods. 2001;255 (1-2):27–40. http://dx.doi.org/10.1016/S0022-1759(01)00535-X.

45. Olson WC, Smolkin ME, Farris EM, Fink RJ, Czarkowski AR, Fink JH, et al. Shipping blood to a central laboratory in multicenter clinical trials: effect of ambient temperature on specimen temperature, and effects of temperature on mononuclear cell yield, viability and immunologic function. J Transl Med. 2011;9:26. doi:10.1186/1479-5876-9-26.

46. Watkins SK, Zhu Z, Watkins KE, Hurwitz AA. Isolation of immune cells from primary tumors. J Vis Exp. 2012;64, e3952. doi:10.3791/3952.

47. Russell ND, Hudgens MG, Ha R, Havenar-Daughton C, McElrath MJ. Moving to HIV-1 vaccine efficacy trials: defining T cell responses as potential correlates of immunity. J Infect Dis. 2003;187:226–42. doi:10.1086/367702.

48. Smith JG, Liu X, Kaufhold RM, Clair J, Caulfield MJ. Development and validation of a gamma interferon ELISPOT assay for quantitation of cellular immune responses to varicella-zoster virus. Clin Diagn Lab Immunol. 2001;8(5):871–9. doi:10.1128/CDLI.8.5.871-879.2001.

49. Weinberg A, Song LY, Wilkening C, Sevin A, Blais B, Louzao R, et al. Optimization and limitations of use of cryopreserved peripheral blood mononuclear cells for functional and phenotypic T-cell characterization. Clin Vaccine Immunol. 2009;16(8):1176–86. doi:10.1128/CVI.00342-08.

50. Yokoyama WM, Thompson ML, Ehrhardt RO. Cryopreservation and thawing of cells. Curr Protoc Immunol. 2012;Appendix 3:3G. doi:10.1002/0471142735.ima03gs99.

51. Filbert H, Attig S, Bidmon N, Renard BY, Janetzki S, Sahin U, et al. Serum-free freezing media support high cell quality and excellent ELISPOT assay performance across a wide variety of different assay protocols. Cancer Immunol Immunother. 2013;62(4):615–27. doi:10.1007/s00262-012-1359-5.

52. Disis ML, dela Rosa C, Goodell V, Kuan LY, Chang JC, Kuus-Reichel K, et al. Maximizing the retention of antigen specific lymphocyte function after cryopreservation. J Immunol Methods. 2006;308(1-2):13–8. doi:10.1016/j.jim.2005.09.011.

53. Weinberg A, Song LY, Wilkening CL, Fenton T, Hural J, Louzao R, et al. Optimization of storage and shipment of cryopreserved peripheral blood mononuclear cells from HIV-infected and uninfected individuals for ELISPOT assays. J Immunol Methods. 2010;363(1):42–50. doi:10.1016/j.jim.2010.09.032.

54. Owen RE, Sinclair E, Emu B, Heitman JW, Hirschkorn DF, Epling CL, et al. Loss of T cell responses following long-term cryopreservation. J Immunol Methods. 2007;326(1-2):93–115. doi:10.1016/j.jim.2007.07.012.

55. Bissoyi A, Nayak B, Pramanik K, Sarangi SK. Targeting cryopreservation-induced cell death: a review. Biopreserv Biobank. 2014;12(1):23–34. doi:10.1089/bio.2013.0032.

56. Gad E, Rastetter L, Herendeen D, Curtis B, Slota M, Koehniein M, et al. Optimizing the cryopreservation of murine splenocytes for improved antigen-specific T cell function in ELISPOT. J Immnother Cancer. 2013;1(Suppl1):211. doi:10.1186/2051-1426-1-S1-P211.

57. Mossoba ME, Walia JS, Rasaiah VI, Buxhoeveden N, Head R, Ying C, et al. Tumor protection following vaccination with low doses of lentivirally transduced DCs expressing the self-antigen erbB2. Mol Ther. 2008;16(3):607–17. doi:10.1038/sj.mt.6300390.

58. Strober W. Trypan blue exclusion test of cell viability. Curr Protoc Immunol. 2001;Appendix 3:Appendix 3B. doi:10.1002/0471142735.ima03bs21.

59. Louis KS, Siegel AC. Cell viability analysis using trypan blue: manual and automated methods. Methods Mol Biol. 2011;740:7–12. doi:10.1007/978-1-61779-108-6_2.

60. Mascotti K, McCullough J, Burger SR. HPC viability measurement: trypan blue versus acridine orange and propidium iodide. Transfusion. 2000;40(6):693–6. doi:10.1046/j.1537-2995.2000.40060693.x.

61. Altman SA, Randers L, Rao G. Comparison of trypan blue dye exclusion and fluorometric assays for mammalian cell viability determinations. Biotechnol Prog. 1993;9(6):671–4. doi:10.1021/bp00024a017.

62. Lenders K, Ogunjimi B, Beutels P, Hens N, Van Damme P, Berneman ZN, et al. The effect of apoptotic cells on virus-specific immune responses detected using IFN-gamma ELISPOT. J Immunol Methods. 2010;357(1-2):51–4. doi:10.1016/j.jim.2010.03.001.

63. Smith JG, Joseph HR, Green T, Field JA, Wooters M, Kaufhold RM, et al. Establishing acceptance criteria for cell-mediated-immunity assays using frozen peripheral blood mononuclear cells stored under optimal and suboptimal conditions. Clin Vaccine Immunol. 2007;14(5):527–37. doi:10.1128/CVI.00435-06.

64. Janetzki S, Britten CM. The impact of harmonization on ELISPOT assay performance. Methods Mol Biol. 2012;792:25–36. doi:10.1007/978-1-61779-325-7_2.

65. Britten CM, Gouttefangeas C, Welters MJ, Pawelec G, Koch S, Ottensmeier C, et al. The CIMT-monitoring panel: a two-step approach to harmonize the enumeration of antigen-specific CD8+ T lymphocytes by structural and functional assays. Cancer Immunol Immunother. 2008;57(3):289–302. doi:10.1007/s00262-007-0378-0.

66. Kutscher S, Dembek CJ, Deckert S, Russo C, Korber N, Bogner JR, et al. Overnight resting of PBMC changes functional signatures of antigen specific T- cell responses: impact for immune monitoring within clinical trials. PLoS One. 2013;8(10), e76215. doi:10.1371/journal.pone.0076215.

67. Römer PS, Berr S, Avota E, Na SY, Battaglia M, ten Berge I, et al. Preculture of PBMCs at high cell density increases sensitivity of T-cell responses, revealing cytokine release by CD28 super-agonist TGN1412. Blood. 2011;118(26):6772–82. doi:10.1182/blood-2010-12-319780.

68. Wegner J, Hackenberg S, Scholz CJ, Chuvpilo S, Tyrsin D, Matskevich AA, et al. High-density preculture of PBMCs restores defective sensitivity of circulating CD8 T cells to virus- and tumor-derived antigens. Blood. 2015;126(2):185–94. doi:10.1182/blood-2015-01-622704.

69. Stefanova I, Dorfman JR, Germain RN. Self-recognition promotes the foreign antigen sensitivity of naive T lymphocytes. Nature. 2002;420(6914):429–34. doi:10.1038/nature01146.

70. Garcia KC, Adams JJ, Feng D, Ely LK. The molecular basis of TCR germline bias for MHC is surprisingly simple. Nat Immunol. 2009;10(2):143–7. doi:10.1038/ni.f.219.

71. Santos R, Buying A, Sabri N, Yu J, Gringeri A, Bender J, et al. Improvement of IFNg ELISPOT performance following overnight resting of frozen PBMC samples confirmed through rigorous statistical analysis. Cells. 2014;4(1):1–18. doi:10.3390/cells4010001.

72. Mata MM, Mahmood F, Sowell RT, Baum LL. Effects of cryopreservation on effector cells for antibody dependent cell-mediated cytotoxicity (ADCC) and natural killer (NK) cell activity in (51)Cr-release and CD107a assays. J Immunol Methods. 2014;406:1–9. doi:10.1016/j.jim.2014.01.017.

73. Helms T, Boehm BO, Asaad RJ, Trezza RP, Lehmann PV, Tary-Lehmann M. Direct visualization of cytokine-producing recall antigen-specific CD4 memory T cells in healthy individuals and HIV patients. J Immunol. 2000;164(7):3723–32. doi:10.4049/jimmunol.164.7.3723.

74. Chaux P, Vantomme V, Coulie P, Boon T, van der Bruggen P. Estimation of the frequencies of anti-MAGE-3 cytolytic T-lymphocyte precursors in blood from individuals without cancer. Int J Cancer. 1998;77(4):538–42. doi:10.1002/(SICI)1097-0215(19980812)77:4<538::AID-IJC11>3.0.CO;2-2.

75. Meidenbauer N, Harris DT, Spitler LE, Whiteside TL. Generation of PSA-reactive effector cells after vaccination with a PSA-based vaccine in patients with prostate cancer. Prostate 2000;43(2):88–100.

76. Chudley L, McCann KJ, Coleman A, Cazaly AM, Bidmon N, Britten CM, et al. Harmonisation of short-term in vitro culture for the expansion of antigen-specific CD8(+) T cells with detection by ELISPOT and HLA-multimer staining. Cancer Immunol Immunother. 2014;63(11):1199–211. doi:10.1007/s00262-014-1593-0.

77. Calarota SA, Baldanti F. Enumeration and characterization of human memory T cells by enzyme-linked immunospot assays. Clin Dev Immunol. 2013;2013:637649. doi:10.1155/2013/637649.

78. Bidmon N, Attig S, Rae R, Schroder H, Omokoko TA, Simon P, et al. Generation of TCR-engineered T cells and their use to control the performance of T cell assays. J Immunol. 2015;194(12):6177–89. doi:10.4049/jimmunol.1400958.

79. Janetzki S, Rabin R. Enzyme-linked ImmunoSpot (ELISpot) for single cell analysis. Methods Mol Biol. 2015;1346:27–46. doi:10.1007/978-1-4939-2987-0_3.

80. Shafer-Weaver K, Rosenberg S, Strobl S, Gregory Alvord W, Baseler M, Malyguine A. Application of the granzyme B ELISPOT assay for monitoring cancer vaccine trials. J Immunother. 2006;29(3):328–35. doi:10.1097/01.cji.0000203079.35612.c8.

81. Falk K, Rotzschke O, Stevanovic S, Jung G, Rammensee HG. Allele-specific motifs revealed by sequencing of self-peptides eluted from MHC molecules. Nature. 1991;351(6324):290–6. doi:10.1038/351290a0.

82. Rammensee HG, Falk K, Rotzschke O. MHC molecules as peptide receptors. Curr Opin Immunol. 1993;5(1):35–44.

83. Rotzschke O, Falk K, Deres K, Schild H, Norda M, Metzger J, et al. Isolation and analysis of naturally processed viral peptides as recognized by cytotoxic T cells. Nature. 1990;348(6298):252–4. doi:10.1038/348252a0.

84. Schumacher TN, Heemels MT, Neefjes JJ, Kast WM, Melief CJ, Ploegh HL. Direct binding of peptide to empty MHC class I molecules on intact cells and in vitro. Cell. 1990;62(3): 563–7. doi:http://dx.doi.org/10.1016/0092-8674(90)90020-F.

85. Ljunggren HG, Stam NJ, Ohlen C, Neefjes JJ, Hoglund P, Heemels MT, et al. Empty MHC class I molecules come out in the cold. Nature. 1990;346(6283):476–80. doi:10.1038/346476a0.

86. Sabbatini P, Tsuji T, Ferran L, Ritter E, Sedrak C, Tuballes K, et al. Phase I trial of overlapping long peptides from a tumor self-antigen and poly-ICLC shows rapid induction of integrated immune response in ovarian cancer patients. Clin Cancer Res. 2012;18(23):6497–508. doi:10.1158/1078-0432.CCR-12-2189.

87. Singh SK, Meyering M, Ramwadhdoebe TH, Stynenbosch LF, Redeker A, Kuppen PJ, et al. The simultaneous ex vivo detection of low-frequency antigen-specific CD4+ and CD8+ T-cell responses using overlapping peptide pools. Cancer Immunol Immunother. 2012;61(11):1953–63. doi:10.1007/s00262-012-1251-3.

88. Kern F, Faulhaber N, Frommel C, Khatamzas E, Prosch S, Schonemann C, et al. Analysis of CD8 T cell reactivity to cytomegalovirus using protein-spanning pools of overlapping pentadecapeptides. Eur J Immunol. 2000;30(6):1676–82. doi:10.1002/1521-4141(200006)30: 6<1676::AID-IMMU1676>3.0.CO;2-V.

89. Russell ND, Hudgens MG, Ha R, Havenar-Daughton C, McElrath MJ. Moving to human immunodeficiency virus type 1 vaccine efficacy trials: defining T cell responses as potential correlates of immunity. J Infect Dis. 2003;187(2):226–42. doi:10.1086/367702.

90. Fiore-Gartland A, Manso BA, Friedrich DP, Gabriel EE, Finak G, Moodie Z, et al. Pooled-peptide epitope mapping strategies are efficient and highly sensitive: an evaluation of methods for identifying human T cell epitope specificities in large-scale HIV vaccine efficacy trials. PLoS One. 2016;11(2):e0147812. doi:10.1371/journal.pone.0147812.

91. de Beukelaar JW, Gratama JW, Smitt PA, Verjans GM, Kraan J, Luider TM, et al. The impact of impurities in synthetic peptides on the outcome of T-cell stimulation assays. Rapid Commun Mass Spectrom. 2007;21(7):1282–8. doi:10.1002/rcm.2958.

92. Currier JR, Galley LM, Wenschuh H, Morafo V, Ratto-Kim S, Gray CM, et al. Peptide impurities in commercial synthetic peptides and their implications for vaccine trial assessment. Clin Vaccine Immunol. 2008;15(2):267–76. doi:10.1128/CVI.00284-07.

93. Suneetha PV, Schlaphoff V, Wang C, Stegmann KA, Fytili P, Sarin SK, et al. Effect of peptide pools on effector functions of antigen-specific CD8+ T cells. J Immunol Methods. 2009;342(1-2):33–48. doi:10.1016/j.jim.2008.11.020.

94. Schmittel A, Keilholz U, Bauer S, Kuhne U, Stevanovic S, Thiel E, et al. Application of the IFN-gamma ELISPOT assay to quantify T cell responses against proteins. J Immunol Methods. 2001;247(1-2):17–24. doi:http://dx.doi.org/10.1016/S0022-1759(00)00305-7.

95. Reyes D, Salazar L, Espinoza E, Pereda C, Castellon E, Valdevenito R, et al. Tumour cell lysate-loaded dendritic cell vaccine induces biochemical and memory immune response in castration-resistant prostate cancer patients. Br J Cancer. 2013;109(6):1488–97. doi:10.1038/bjc.2013.494.

96. Alfaro C, Perez-Gracia JL, Suarez N, Rodriguez J, Fernandez de Sanmamed M, Sangro B, et al. Pilot clinical trial of type 1 dendritic cells loaded with autologous tumor lysates combined with GM-CSF, pegylated IFN, and cyclophosphamide for metastatic cancer patients. J Immunol. 2011;187(11):6130–42. doi:10.4049/jimmunol.1102209.

97. Janetzki S, Song P, Gupta V, Lewis JJ, Houghton AN. Insect cells as HLA-restricted antigen-presenting cells for the IFN-gamma elispot assay. J Immunol Methods. 2000;234(1-2):1–12. doi:http://dx.doi.org/10.1016/S0022-1759(99)00203-3.

98. Henderson RA, Michel H, Sakaguchi K, Shabanowitz J, Appella E, Hunt DF, et al. HLA-A2.1-associated peptides from a mutant cell line: a second pathway of antigen presentation. Science. 1992;255(5049):1264–6. doi:10.1126/science.1546329.

99. Zweerink HJ, Gammon MC, Utz U, Sauma SY, Harrer T, Hawkins JC, et al. Presentation of endogenous peptides to MHC class I-restricted cytotoxic T lymphocytes in transport deletion mutant T2 cells. J Immunol. 1993;150(5):1763–71.

100. Steinle A, Schendel DJ. HLA class I alleles of LCL 721 and 174 x CEM. T2(T2). Tissue Antigens. 1994;44(4):268–70. doi:10.1111/j.1399-0039.1994.tb02394.x.
101. Salcedo M, Momburg F, Hammerling GJ, Ljunggren HG. Resistance to natural killer cell lysis conferred by TAP1/2 genes in human antigen-processing mutant cells. J Immunol. 1994;152(4):1702–8.
102. Zemmour J, Little AM, Schendel DJ, Parham P. The HLA-A,B "negative" mutant cell line C1R expresses a novel HLA-B35 allele, which also has a point mutation in the translation initiation codon. J Immunol. 1992;148(6):1941–8.
103. Janetzki S, Palla D, Rosenhauer V, Lochs H, Lewis JJ, Srivastava PK. Immunization of cancer patients with autologous cancer-derived heat shock protein gp96 preparations: a pilot study. Int J Cancer. 2000;88(2):232–8. doi:10.1002/1097-0215(20001015)88: 2<232::AID-IJC14>3.0.CO;2-8.
104. Maki RG, Livingston PO, Lewis JJ, Janetzki S, Klimstra D, Desantis D, et al. A phase I pilot study of autologous heat shock protein vaccine HSPPC-96 in patients with resected pancreatic adenocarcinoma. Dig Dis Sci. 2007;52(8):1964–72. doi:10.1007/ s10620-006-9205-2.
105. Sinclair NR, Anderson CC. Co-stimulation and co-inhibition: equal partners in regulation. Scand J Immunol. 1996;43(6):597–603. doi:10.1046/j.1365-3083.1996.d01-267.x.
106. Moodie Z, Price L, Gouttefangeas C, Mander A, Janetzki S, Lower M, et al. Response definition criteria for ELISPOT assays revisited. Cancer Immunol Immunother. 2010;59(10):1489–501. doi:10.1007/s00262-010-0875-4.
107. Moodie Z, Huang Y, Gu L, Hural J, Self SG. Statistical positivity criteria for the analysis of ELISpot assay data in HIV-1 vaccine trials. J Immunol Methods. 2006;315(1-2):121–32. doi:10.1016/j.jim.2006.07.015.
108. Dubey S, Clair J, Fu TM, Guan L, Long R, Mogg R, et al. Detection of HIV vaccine-induced cell-mediated immunity in HIV seronegative clinical trial participants using an optimized and validated enzyme-linked immunospot assay. J Acquir Immune Defic Syndr. 2007;45(1):20–7. doi:10.1097/QAI.0b013e3180377b5b.
109. Moodie Z, Price L, Janetzki S, Britten CM. Response determination criteria for ELISPOT: toward a standard that can be applied across laboratories. Methods Mol Biol. 2012;792:185–96. doi:10.1007/978-1-61779-325 7_15.
110. Santos RS, Gringeri A, Yu J, Janetzki S, Judkowski VA, Pinilla C. IMCT-17 Statistically significant association of glioblastoma immunotherapy phase II clinical study (ICT107) treatment and survival to immune response using a novel comprehensive ELISPOT analysis. Neuro-Oncol. 2015;17 Suppl 5:v111. doi:10.1093/neuonc/nov218.17.
111. Santos RG, Bunying A, Sabri N, Yu JS, Swanson S, Judkowski VA et al. A novel methodology to categorize immune responders in cancer immunotherapy trials based on fusion of response metrics and Monte Carlo simulation. Submitted for publication. 2016.
112. Janetzki S, Britten CM, Kalos M, Levitsky HI, Maecker HT, Melief CJM, et al. "MIATA"-minimal information about T cell assays. Immunity. 2009;31:527–8. doi:10.1016/j. immuni.2009.09.007.
113. Hoos A, Eggermont AM, Janetzki S, Hodi FS, Ibrahim R, Anderson A, et al. Improved endpoints for cancer immunotherapy trials. J Natl Cancer Inst. 2010;102(18):1388–97. doi:10.1093/jnci/djq310.
114. van der Burg SH, Kalos M, Gouttefangeas C, Janetzki S, Ottensmeier C, Welters MJ, et al. Harmonization of immune biomarker assays for clinical studies. Sci Transl Med. 2011;3(108):108ps44. doi:10.1126/scitranslmed.3002785.
115. Maecker HT, Hassler J, Payne JK, Summers A, Comatas K, Ghanayem M, et al. Precision and linearity targets for validation of an IFNgamma ELISPOT, cytokine flow cytometry, and tetramer assay using CMV peptides. BMC Immunol. 2008;9:9. doi:10.1186/1471-2172-9-9.
116. Taylor CF, Field D, Sansone S-A, Aerts J, Apweiler R, Ashburner M, et al. Promoting coherent minimum reporting guidelines for biological and biomedical investigations: the MIBBI project. Nat Biotechnol. 2008;26(18688244):889–96. doi:10.1038/nbt.1411.

117. Brazma A, Hingamp P, Quackenbush J, Sherlock G, Spellman P, Stoeckert C, et al. Minimum information about a microarray experiment (MIAME)-toward standards for microarray data. Nat Genet. 2001;29(4):365–71. doi:10.1038/ng1201-365.
118. Brazma A, Robinson A, Cameron G, Ashburner M. One-stop shop for microarray data. Nature. 2000;403(6771):699–700. doi:10.1038/35001676.
119. Lee JA, Spidlen J, Boyce K, Cai J, Crosbie N, Dalphin M, et al. MIFlowCyt: the minimum information about a Flow Cytometry Experiment. Cytometry A. 2008;73(10):926–30. doi:10.1002/cyto.a.20623.
120. Britten CM, Janetzki S, van der Burg SH, Huber C, Kalos M, Levitsky HI, et al. Minimal information about T cell assays: the process of reaching the community of T cell immunologists in cancer and beyond. Cancer Immunol Immunother. 2011;60(1):15–22. doi:10.1007/s00262-010-0940-z.
121. Britten CM, Janetzki S, Butterfield LH, Ferrari G, Gouttefangeas C, Huber C, et al. T cell assays and MIATA: the essential minimum for maximum impact. Immunity. 2012;37(1):1–2. doi:10.1016/j.immuni.2012.07.010.

Index

© Springer International Publishing Switzerland 2016
S. Janetzki, *Elispot for Rookies (and Experts Too)*, Techniques in Life Science
and Biomedicine for the Non-Expert, DOI 10.1007/978-3-319-45295-1

Printed in the United States
By Bookmasters